STARWATCH

STARWATCH

A MONTH-BY-MONTH GUIDE TO THE NIGHT SKY

ROBIN KERROD

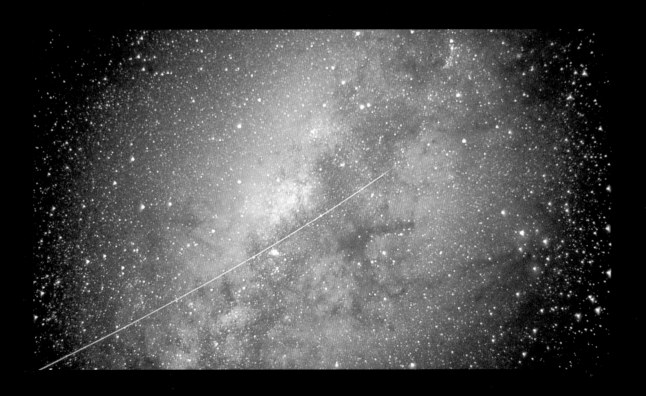

BARRON'S

A QUARTO BOOK

First edition for the United States, its territories and dependencies, and Canada published in 2003 by Barron's Educational Series, Inc.

All inquiries should be addressed to:
Barron's Educational Series, Inc.
250 Wireless Boulevard
Hauppauge, New York 11788
http://barronseduc.com

Copyright © 2003 Quarto Inc.

All rights reserved. No part of this book may be reproduced in any form, by photostat, microfilm, xerography, or any other means, or incorporated into any information retrieval system, electronic or mechanical, without the written permission of the copyright owner.

International Standard Book No. 0-7641-5666-7
Library of Congress Catalog Card No. 2003108751

QUAR.TYA

Conceived, designed, and produced by
Quarto Publishing plc
The Old Brewery
6 Blundell Street
London N7 9BH

Project Editor: Kate Tuckett
Senior Art Editor: Sally Bond
Designer: Tanya Devonshire Jones
Photographer: Colin Bowling
Illustrators: Richard Monkhouse,
              John Cox, Sally Bond
Copy Editor: Rod Cuff
Proofreader: Anne Plume
Indexer: Diana Le Core

Art Director: Moira Clinch
Publisher: Piers Spence

Manufactured by Universal Graphics Pte Ltd., Singapore
Printed by Midas Printing International Ltd., China

9 8 7 6 5 4 3 2 1

CONTENTS

INTRODUCTION	6

GETTING STARTED 8

TOOLS OF THE TRADE	10
THE FAMILY OF THE SUN	12
LOOKING AT THE MOON	14
PROFILING—THE MOON	16
LOOKING AT PLANETS	18
PROFILING—THE PLANETS	20
LOOKING AT COMETS	22
LOOKING AT STARS	24
PROFILING—THE STARS	26

THE SKIES MONTH BY MONTH 28

JANUARY SKIES 30
   FROM AURIGA TO ORION 32
   NEBULAS—CLOUDS AMONG THE STARS 34

FEBRUARY SKIES 36
   FROM CANCER TO GEMINI 38

MARCH SKIES 40
   FROM CARINA TO VELA 42
   KEY CONSTELLATIONS—SIGNPOSTS TO THE STARS 44

APRIL SKIES 46
   FROM CANES VENATICI TO VIRGO 48

MAY SKIES 50
   FROM BOÖTES TO URSA MINOR 52
   THE ZODIAC—THE CIRCLE OF ANIMALS 54

JUNE SKIES 56
   FROM CEPHEUS TO SCORPIUS 58

JULY SKIES 60
   FROM ARA TO SERPENS 62
   THE MILKY WAY—THE ARCH OF COUNTLESS STARS 64

AUGUST SKIES 66
   FROM AQUILA TO SAGITTARIUS 68
   METEORS—STARS THAT FALL FROM THE SKY 70

SEPTEMBER SKIES 72
   FROM AQUARIUS TO PISCIS AUSTRINUS 74

OCTOBER SKIES 76
   FROM ANDROMEDA TO TUCANA 78
   GALAXIES—STAR ISLANDS IN SPACE 80

NOVEMBER SKIES 82
   FROM ARIES TO TRIANGULUM 84

DECEMBER SKIES 86
   FROM COLUMBA TO TRIANGULUM AUSTRALE 88
   CLUSTERS—STARS THAT TRAVEL TOGETHER 90

WORDS TO REMEMBER 92
INDEX 94
CREDITS 96

INTRODUCTION

Every night (when the weather is clear) you can see thousands of stars shining down. They sparkle like jewels scattered on black velvet—the black velvet of deep space. The brightest stars make patterns in the sky that we can learn to recognize. We call these star patterns "constellations." They help us find our way across the night sky.

The night sky is changing all the time. The constellations change their positions as the nights go by. And, month by month, new constellations appear while others disappear. All these changes take place regularly at the same times every year. So we can draw up star maps that show how the night sky looks every month. Using the *Starwatch* maps beginning on page 30, you can learn about the constellations highlighted each month and about featured subjects such as nebulas and the zodiac.

When you begin stargazing and noting the changes that take place in the night sky, you will be following one of the most popular pastimes in the world—astronomy. Astronomy is the scientific study of the heavens. It is perhaps the most ancient of the sciences, which began in the early civilizations of the Middle East more than 5,000 years ago.

But astronomy is not just about looking at stars. It is about studying all the strange things that happen in the night sky, such as the wanderings of the planets, the coming and going of comets, and the unforgettable spectacle of eclipses. Welcome to the fascinating world of astronomy!

RIGHT All the stars in the Universe are gathered together in great star islands in space, which we call *galaxies*. This galaxy is found in the constellation Triangulum (the Triangle).

GETTING STARTED

To get started in astronomy, you don't need any equipment at all—just your eyes. And you can see a great deal with just your eyes, or "with the naked eye." You can see: The patterns of stars we call the constellations, the bright "wandering stars" we call the planets, the ever-changing Moon, the fiery streaks we call falling stars, and sometimes bright comets with long tails.

ABOVE When viewing eclipses, use glasses that filter out the Sun's rays.

BE PREPARED

It's always worth planning ahead before you go stargazing. To begin with, make sure you are dressed properly. Clear nights that are good for stargazing are often the coldest. So wrap up well: In winter wear a sweater and an anorak, with a woolly hat, gloves, warm socks, and waterproof boots. In summer, you can wear lighter clothing, but keep a sweater handy because even summer nights can feel cold, especially if you're standing or sitting still for long periods.

Take a stool or even a deckchair to sit on, and a foldaway camping table to put things on. Also, take a hot drink and a snack if you intend stargazing for some time.

DON'T FORGET

You'll find stargazing more rewarding if you take along a few essential items. One is a notebook with pen or pencil, so you can make notes about what you see. You may want to make sketches, too.

Take a watch to note the time when you make your observations. You will also need to know the direction of your observations. So take a compass, unless you already know your bearings (directions). (In the Northern Hemisphere, you can get your bearings by finding the Pole Star, which is always due north—see page 53.)

LEFT Wear warm clothes when you go stargazing. Nights can get chilly, even during the summer. Hot drinks from a thermos will also help keep you warm.

BELOW Some of the other bits and pieces you'll find useful when you go stargazing. Cover the torch with red film because red light doesn't affect your night vision.

GETTING STARTED 009

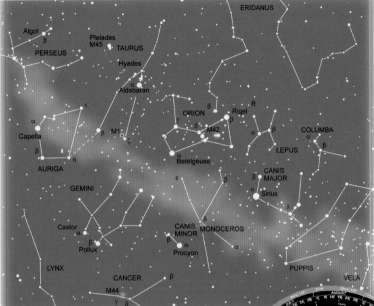

SKY MAPS

One thing you'll always need when you go stargazing is a set of star maps, like the ones in this book (see page 30 onward). These maps will show you the constellations that you can see in each month.

A planisphere is very useful too. One comes with this book. You'll find it in a pocket at the back. This handy device shows you views of the night sky for every night of the year. You match the time of observation with the day of the month on the two disks, and then the stars on view appear in the window.

To read the maps and planisphere and to write your notes, use a red light. Red light doesn't affect your night vision; white light does. You can buy special red lights or cover an ordinary flashlight with red film.

ABOVE Use star maps like the ones in this book to help find your way around the heavens every month.

ABOVE A planisphere gives you a whole-sky view of the heavens for every night of the year.

NIGHT VISION

Before you start stargazing, you must let your eyes get used to the dark. This can take half an hour or longer. During this time, the pupils of the eye open wider to let more light in, and the retina—the "screen" at the back of the eye—becomes more sensitive to light. So when you eventually get your night vision, you will be able to see many more stars than when you started.

TOOLS OF THE TRADE

You can see many things in the night sky with the naked eye. But with a little help, you can see a great deal more. The most useful tools for amateur astronomers are a pair of binoculars and a small telescope. They gather much more light than the eye does, and provide stunning views of colorful stars, billowing nebulas (gas clouds), dazzling star clusters, and faraway galaxies.

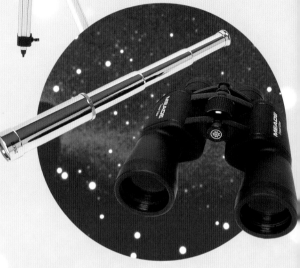

RIGHT The Hubble Space Telescope sends back beautiful pictures of the Universe.

LEFT AND BELOW Telescopes and binoculars are the astronomer's most useful tools.

USING LENSES

The Italian astronomer Galileo first built a telescope to look at the sky in 1609. He made it using glass lenses. Today many astronomers still use lens telescopes, which are called refractors.

In a refractor, one lens gathers starlight and forms an image. You look at the image through another lens, called the eyepiece.

Binoculars also use lenses to gather and focus light. They are a kind of compact double telescope.

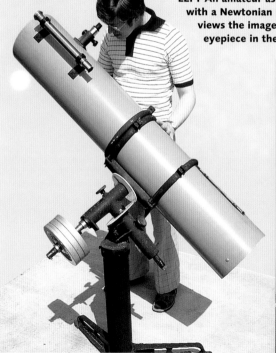

LEFT An amateur astronomer with a Newtonian reflector. He views the image through an eyepiece in the side.

BELOW Inside these domes at Kitt Peak National Observatory in Arizona are powerful telescopes that look deep into space.

TOOLS OF THE TRADE

USING MIRRORS

However, most astronomers use a telescope called a reflector, which has mirrors to gather and focus light. In a reflector, a curved main mirror gathers light and reflects it to another mirror. This mirror in turn reflects the light into an eyepiece.

The most powerful telescopes on Earth have huge mirrors. The twin Keck telescopes on Hawaii have mirrors 33 feet (10 m) across. They give us magnificent views of the sky. So does the Hubble Space Telescope in orbit around the Earth. It sees much more clearly because it circles high above Earth's dirty atmosphere.

FOCUS ON PHOTOGRAPHY

If you follow a few simple rules, there is nothing difficult about astrophotography, or taking photographs of stars. But don't expect to get the magnificent pictures you see in astronomy books and magazines!

All you need is a good camera that has a B setting—this allows you to keep the shutter open for a long time. You need this because there is little light at night, and you need to let it build up on the film. A good all-around film to use is ISO 400.

You will also need a tripod on which to mount the camera, and a cable shutter-release. The tripod holds the camera steady, and the cable release lets you open and close the shutter without touching—and wobbling—the camera.

To capture star trails, just point the camera to any part of the sky and open the shutter. Leave the shutter open for up to an hour, and this will give you some nice trails. To photograph constellations, you should use a faster film and shorter exposures of a minute or so. Then the stars will show up as dots rather than as trails.

ABOVE Point your camera at the night sky and open the shutter for a while. You'll capture beautiful star trails as the heavens revolve overhead.

BELOW Essential equipment for astrophotography—a camera with a B setting, a shutter release, and a tripod.

THE FAMILY OF THE SUN

Of all the stars in the sky, we can single out one as being very special. But it is a star of the day, not of the night. It is the daytime star we know better as the Sun. The Sun is an ordinary star, just like the thousands we can see in the night sky. It looks very much bigger and brighter than the nighttime stars only because it is very much closer to us.

ABOVE One of the "miniplanets" found in the asteroid belt. It's called Ida.

FAMILY OF PLANETS

From Earth, we see the Sun rise every morning in the east, travel across the sky, then set in the evening in the west. It seems as if the Sun is circling around Earth. In fact, it's the other way around—the Earth travels around the Sun. It is a body we call a planet.

Eight other planets circle around the Sun, making up the most important part of the Sun's family, or Solar System. We can see five of the other planets in the night sky. They look like really bright stars. Read more about the planets on page 18.

OTHER FAMILY MEMBERS

Many other smaller bodies belong to the Solar System. To us, the most important one is the Moon, which shines down on us most nights. The Moon circles in space around Earth about once a month. Read more about the Moon on page 14. Most other planets have moons too—altogether there are more than 100 moons.

Swarms of other bodies are found in the Solar System too. There are many thousands of miniplanets, which we call the asteroids. We can't see them in the night sky because they are too small and too far away. But we can see even smaller icy lumps when they travel in toward the Sun and start to melt—we see them as comets. Find out more about comets on page 22.

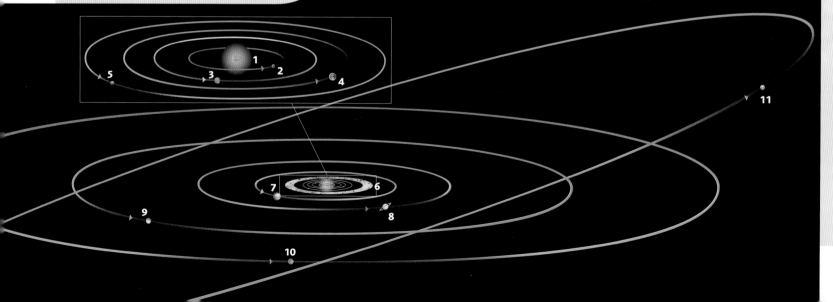

BELOW The planets move in circles around the Sun, all in the same direction. The inner planets are quite close together. The outer planets are widely separated. Most of the Solar System consists of empty space.

1 Sun
2 Mercury
3 Venus
4 Earth
5 Mars
6 Asteroid belt
7 Jupiter
8 Saturn
9 Uranus
10 Neptune
11 Pluto

THE FAMILY OF THE SUN 013

RIGHT Violent activity in the Sun's atmosphere, captured by the spacecraft SOHO.

BELOW In a total eclipse of the Sun, the Sun's outer atmosphere, or corona, shows up.

PROFILING—THE SUN

The Sun is different from all the other bodies in the Solar System because it is a star, made up of extremely hot gas. On the surface, its temperature is nearly 10,000 degrees Fahrenheit (5,500 degrees Celsius). Inside, the temperature rises to tens of millions of degrees.

From its hot, bright surface, the Sun pours out enormous energy into space as heat, light, and other rays we can't see. These invisible rays include X-rays and ultraviolet rays, which are the rays that give us sunburn. The Sun is the only body in the Solar System that gives off light. The other bodies shine only because they reflect the Sun's light.

The Sun is much bigger than all the other bodies in the Solar System. With a diameter of about 865,000 miles (1,400,000 km), it is over 100 times as wide as Earth.

GOOD THINKING

Until a few hundred years ago, people thought that the Sun, the Moon, and the planets circled around Earth. They believed that Earth was the center of the Universe. But, in 1543, a Polish astronomer named Nicolaus Copernicus suggested that Earth and the planets circled around the Sun. He was right, of course.

ABOVE Copernicus's drawing that depicted planets circling round the Sun.

LOOKING AT THE MOON

The Moon is our nearest neighbor in space, and we can see it in the sky most nights. The Moon circles around Earth about once a month. During this time, it seems to change shape—from a thin sliver to a full circle and back again. These changing shapes are called the phases of the Moon. They mark one of the great rhythms of nature.

NEW MOON

The Moon shines in the night sky, but it does not give out light of its own. We see it shine only because it reflects light from the Sun. As the Moon circles around Earth, we see different amounts of it lit up by the Sun at different times each month. And this is what causes the phases.

The phases begin when the Sun shines only on the opposite or far side of the Moon. The side facing us—the near side—is in darkness, so we can't see the Moon at all. We call this phase the New Moon.

ABOVE A Full Moon, when all of the near side of the Moon is lit up by the Sun.

ABOVE Crescent phase, when the Moon is only a few days old. The circular sea is called the Sea of Crises.

WAXING AND WANING

A few days later, as the Moon moves further along its orbit, we see the edge of the Moon lit up. We call this a Crescent Moon. Gradually, more and more of the Moon gets lit up. After about a week, half is lit up and we call it the First Quarter phase. After another week, the whole face is lit up and we call it the Full Moon.

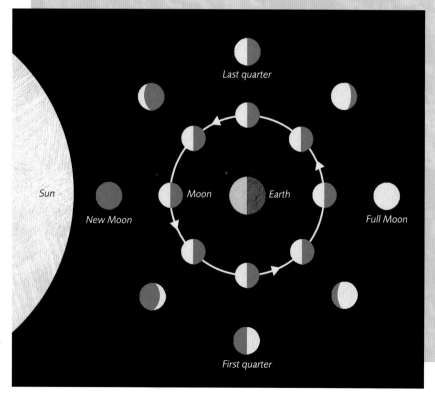

LEFT The progression of the Moon's phases as the Moon circles the Earth. The outer circles show the different shapes of the Moon as we see it.

Up to this stage, we say that the Moon has been waxing, or getting bigger. Now it starts waning, or getting smaller. About a week after Full Moon, only about half the surface is lit up, and we call it the Last Quarter. After another week, only a slim crescent remains lit up. Then the Moon disappears. It's the next New Moon, 29½ days since the last one.

LIGHT AND DARK

The Moon is so close that we can see some features on its surface even with the naked eye. There are dark areas and light areas. Early astronomers thought that the dark areas might be seas and the light ones continents, like the seas and land areas on Earth. They called the dark areas *maria*, which is Latin for "seas." And we still call them this (see page 16).

If you look at the Moon month after month, you notice that its surface always looks the same. In other words, we always see the same side of the Moon—the near side. We never see the other side—the far side.

This happens because of the way the Moon spins around as it travels in its orbit around Earth. It spins round once in exactly the same time as it takes to circle Earth once.

ABOVE An astronaut's view of the Moon. This picture mainly shows the far side of the Moon, invisible from Earth.

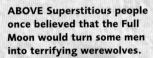

ABOVE Superstitious people once believed that the Full Moon would turn some men into terrifying werewolves.

LUNATICS AND WEREWOLVES

In earlier times, people held many superstitions about the Moon. For example, they believed it was dangerous to stay out in the light of the Full Moon for too long. It would drive you mad, they said. This is how we got our word "lunatic," meaning a mad person, from luna, the Latin word for moon. Another superstition was that the Full Moon's rays would turn some people into vicious, wolflike creatures called werewolves that killed and ate people.

PROFILING—THE MOON

BELOW Craters large and small cover the Moon. Some are made by volcanoes, but most are dug out by meteorites.

Like Earth, the Moon is made up of rock. Measuring only 2,160 miles (3,476 km) across, it is much smaller than Earth. It also has no atmosphere (air). This is because its gravity, or pull, is not strong enough to hang on to any gases to make an atmosphere.

Nevertheless, the Moon's gravity is still strong enough to affect Earth, 239,000 miles (385,000 km) away. It tugs at the water in the oceans and causes the tides.

Without an atmosphere, there is no weather on the Moon as we know it. There are no clouds in the sky, no rain, no frost, no snow, and no rivers of running water. However, space probes have found that there is water on the Moon, locked as ice in deep craters.

LUNAR LANDSCAPES

We now know that the dark areas we see on the Moon are not seas, but are really great dusty plains. We still call them seas, or *maria* (Latin for "seas"). The lighter areas we see are mainly highland areas. In places, the land rises to form long mountain chains, with peaks as high as 65,000 feet (20,000m).

LEFT The colorful Earth rises over the horizon of the drab, rugged Moon.

Strangely, all the main seas are on the near side, the side of the Moon that always faces us. The hidden side is mostly rugged and mountainous with hardly any seas.

The other main features of the lunar landscape are the millions of pits, or craters. Lumps of rock called meteorites dug out these craters when they hit the Moon long ago. There are craters of all sizes—the biggest measure more than 150 miles (250 km) across.

PROFILING—THE MOON

ABOVE A typical moon rock, riddled with holes where gas escaped from it. Like all moon rocks, it is volcanic.

MOON SOIL AND ROCKS

We know a lot about the Moon's surface, thanks to the Apollo astronauts. They took samples of surface soil and rocks and brought them back to Earth to be examined.

The surface is covered with a few inches (cm) of a kind of soil, which scientists call regolith. It is made up of dust, rocky bits, and tiny beads of glass. These form when meteorites smash up the surface rocks.

There are not as many kinds of rock on the Moon as there are on Earth. On the Moon, they are all volcanic rocks, formed when molten rock cools. One of the commonest kinds is like the dark rock we call basalt on Earth.

FOOTSTEPS ON THE MOON

On July 20, 1969, Apollo 11 astronaut Neil Armstrong became the first human to plant his footprints on the Moon. Apollo 11 was the first of six missions that landed on the Moon. The last was in December 1972. During these missions, 12 astronauts explored the surface on foot for over 80 hours and brought back 850 pounds (385 kg) of moon rocks.

RIGHT Apollo 17 astronaut Jack Schmitt examines a huge boulder at the edge of the Moon's Sea of Serenity.

LOOKING AT PLANETS

On many nights, you can see a bright star in the western sky just after the Sun has set. But this evening star is not a star at all—it's the planet Venus. We can also see some of the other planets, usually shining brighter than the stars. Early astronomers noticed that these bright objects changed their position every night, unlike ordinary stars. So they called them planets, a word meaning "wanderers."

ABOVE Mars and Jupiter as seen in the evening sky after sunset. After Venus, they are the brightest of the planets.

EVENING AND MORNING STARS

Venus shines more brightly than any other star or planet. After the Moon, it is the brightest object in the night sky. It appears so bright because it comes closer to us than any other planet, sometimes as close as 26 million miles (42 million km). And it is covered in thick clouds, which reflect sunlight brilliantly.

We see Venus as an evening star for only part of the year. At other times, we may see it in the east before sunrise as a morning star.

Venus is not the only morning and evening star. Mercury can be one, too. It does not become nearly as bright as Venus. It can be difficult to see because it always stays quite close to the Sun and is often lost in the Sun's glare. It never gets very high above the horizon.

PLANETS OF THE NIGHT

Three other planets are visible to the naked eye—Mars, Jupiter, and Saturn. You may sometimes spot them in the twilight, but usually they are seen during the dark skies of night.

After Venus, Mars comes closer to Earth than the other planets. And at times it can become really bright. It becomes its brightest at opposition, when it is opposite the Sun in the sky and nearest to Earth. Among the planets, Mars is unmistakable because it shines with a reddish hue. Ancient astronomers likened the color of Mars to blood and fire, and named it after their god of war.

LEFT Planet Venus shows up in the sky at dawn, when we call it the morning star. At other times we see it after sunset as the evening star.

LOOKING AT PLANETS

ABOVE The planets are all different sizes—from tiny Pluto (1,430 miles or 2,300 kilometers across) to gigantic Jupiter (88,900 miles or 143,000 kilometers across).

THE BRIGHT GIANTS

The giant planet Jupiter often shines as bright as Mars at its brightest. If you see a brilliant white "star" in the night sky, it is sure to be Jupiter. This planet lies very much further away from us than reddish Mars does, but shines so brightly because it is so big—it is over 20 times bigger across than Mars.

Saturn also becomes brighter than most stars when it is closest to us. But it becomes faint and difficult to find among the other stars when it moves far away.

We can't see the other three planets with the naked eye—they are too far away. Uranus, for example, lies twice as far away as Saturn and is also much smaller. Even powerful telescopes tell us little about these remote worlds.

Jupiter

NEW WORLDS

Ancient astronomers thought there were only six planets—the ones they could see with their eyes. English astronomer William Herschel caused a sensation when he found a seventh planet (Uranus) in 1781. German astronomer Johann Galle made it eight when he discovered another (Neptune) in 1846. And U.S. astronomer Clyde Tombaugh discovered planet number nine (Pluto) in 1930.

Neptune

PROFILING—THE PLANETS

The other eight planets are all very different from our home planet Earth. In particular, Earth is the only one on which conditions are just right for life. As far as we know, there is no life on any other planet. There is a possibility that there once might have been life on Mars, because that planet once had a much milder climate than it has today.

We can divide the planets into two main kinds of body The four inner planets are rocky bodies like Earth. So we call them the rocky or terrestrial ("Earthlike") planets. The next four are much bigger and are made up mainly of gas. So we call them gas giants. The one farthest away, Pluto, is different again. It is a tiny ice-world.

THE ROCKY PLANETS

The four rocky planets are Mercury, Venus, Earth, and Mars. Being closest to the Sun, Mercury is scorching hot, with temperatures rising in places as high as 840 degrees Fahrenheit (450 degrees Celsius). Its baked surface is almost completely covered with craters. It has hardly a trace of an atmosphere.

The next planet out, Venus, has a very thick atmosphere and is slightly hotter. Clouds prevent us seeing what its surface is like. But space probes have pictured the surface using radar, which can see through clouds. The surface is almost completely covered with lava from volcanoes.

Mars is farther away from the Sun than Earth is, and is much colder. Clouds float in its thin atmosphere. The surface is covered with reddish rocks and soil. There are great desert regions, and areas pitted with craters. There are also three huge volcanoes and a great canyon system, like Arizona's Grand Canyon but much bigger.

BELOW Our home planet Earth, colored predominantly blue by the oceans and white by clouds and ice.

ABOVE Volcanoes, and the lava they spew, cover most of the surface of Venus.

Mars

RIGHT Red rocks litter the surface of the Red Planet, Mars, in this picture sent back by Pathfinder.

PROFILING THE PLANETS

ABOVE With its system of bright, shining rings, Saturn is the most beautiful planet of them all.

THE GAS GIANTS

The next two planets, Jupiter and Saturn, are truly gigantic. Jupiter is more than ten times as big across as Earth. It looks very colorful because of the reddish bands of clouds speeding through the thick atmosphere of hydrogen and helium.

Underneath the clouds, there is no solid surface, but a great ocean of liquid hydrogen and helium that covers the whole planet. Saturn has a similar makeup, but its main claim to fame is its magnificent system of shining rings.

Uranus and Neptune are the smallest gas giants, and they too have deep oceans underneath the atmosphere. They are each about four times bigger across than Earth.

ICY PLUTO

Tiny Pluto is the odd one out among the planets. It is a deeply frozen world, made up of rock and ice. It is much like Neptune's largest moon, Triton. And astronomers think that they are the biggest of a whole swarm of ice worlds lurking in the depths of the Solar System.

LEFT Giant planet Jupiter with one of its four large moons, Europa. Between them, the gas giants have over 100 moons.

LOOKING AT COMETS

Every now and then, some of the most spectacular of all celestial bodies pass through our skies. They have large glowing heads and long tails fanning out behind. Ancient Chinese astronomers called them broom stars, but they are not stars at all—they are comets. Comets belong to the Solar System and travel in orbit around the Sun.

ABOVE Comet Hale-Bopp shone brightly in northern skies for weeks in early 1997.

EVIL OMENS

For most of the time, when they are far from the Sun, comets can't be seen. They start to shine only when they get near the Sun, and suddenly seem to appear in the sky. Because of this, they used to scare people long ago, who looked for good and bad signs in the sky. To them, a comet was a bad sign, which could bring disease, war, famine, and death.

There is the famous story of the comet that appeared in the skies at the time of the Battle of Hastings in England in 1066. It proved a bad sign for the English: King Harold was killed and then the English were defeated by William the Conqueror.

BELOW Comets grow a bright head (and usually twin tails) as they near the Sun.

HAPPY RETURNS

People didn't realize that comets belonged to the Solar System until the 1600s. Then, English astronomer Edmond Halley showed that the comet he saw in 1681 was a regular visitor to Earth's skies. It was named after him.

It so happens that the 1066 comet was an earlier appearance of Halley's Comet. The comet last returned in 1986 and will be seen again in about 2061. It circles once around the Sun every 76 years or so.

Many of the brightest comets may be traveling toward the Sun for the first time or returning after thousands of years. Hale-Bopp, the brilliant comet of 1997, was probably last seen when the Egyptians were building the Pyramids about 4,000 years ago.

PROFILING—COMETS

What exactly are comets? Mostly, they are great clouds of gas and dust. They have only a tiny solid part measuring just a few miles across. This is called the comet's nucleus.

The nucleus is a lump of rocky bits and dust frozen in ice. It is sometimes called a "dirty snowball." When the comet is far from the Sun, this lump remains deeply frozen, and invisible. When it draws near the Sun, it begins to heat up. The ice melts and then turns to vapor (gas), releasing a lot of dust.

The gas and dust form a great cloud around the nucleus. The cloud reflects sunlight, and the comet becomes visible. Streams of particles coming from the Sun force the cloud away from the nucleus to form the tail.

BELOW Comet Hyakutake was the most spectacular comet of 1996. It grew a particularly long tail.

ABOVE Jets of gas spurt from the nucleus of Halley's Comet when it neared Earth in 1986. The probe Giotto took the picture.

NAMING COMETS

Comet Hale-Bopp is named after the two U.S. astronomers who first sighted it, in 1995. They were Alan Hale and Thomas Bopp. Every night astronomers around the world are searching the skies for new comets, for no one knows when and where one will appear. If you happen to discover a new comet yourself, you will achieve everlasting fame, because the comet will be named after you.

LOOKING AT STARS

Gaze up at the night sky on a clear night and you'll see thousands of twinkling stars stretching from horizon to horizon. If you were patient, you could probably count two thousand or more. There are many more stars, of course, but they are too faint to be seen with the naked eye. Through a telescope, you can see stars in their millions.

STAR PATTERNS

To begin with, you might find the starry night sky very confusing. There are so many stars.

However, if you stargaze for a while, you notice that some stars are brighter than others and seem to form patterns. If you look in the same part of the sky at about the same time for a few nights running, you will see the same star patterns.

Look at the night sky a year or even ten years later, and you'll always see the same star patterns. We call them the constellations. They help us find our way around the night sky.

ABOVE A map showing the constellations of the Northern Hemisphere from an early star atlas.

GODS AND BEASTS

More than 2,000 years ago, ancient astronomers observed the same constellations that we see today. They imagined they could see in these patterns figures of gods, heroes, animals, and monsters that played a prominent part in their traditional stories, or myths.

Here was their daring hero Perseus, there the princess Andromeda he saved from the savage sea monster Cetus; here a crouching lion, there a flying swan.

We still use the names that ancient astronomers gave the constellations, in their Latin forms. So the lion is Leo, and the swan is Cygnus. These two constellations look quite like the figures they are meant to be, but most constellations don't.

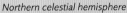

LEFT The imaginary celestial sphere that appears to surround the Earth.

Path of Sun
Celestial equator

OUR BOUNDLESS UNIVERSE

When we look up into the starry sky, we are looking out at one little corner of our Universe. The Universe is everything that exists—stars and space, energy and forces. The stars we see in our skies are just the "tip of the iceberg" as far as the Universe is concerned. They belong to a great star island in space, a galaxy, containing hundreds of billions of stars. And there are billions of galaxies like it in the Universe.

BELOW The constellations of the northern (left) and southern celestial hemispheres.

Northern celestial hemisphere

Southern celestial hemisphere

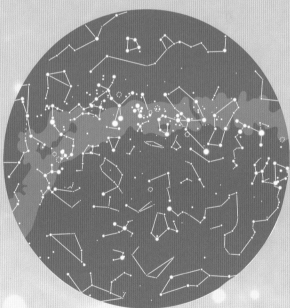

THE CELESTIAL SPHERE

Long ago, people imagined that the stars were stuck on the inside of a great dark dome above their heads. They called it the celestial sphere. We know now that this isn't true. The stars are not all the same distance away, but are spread out over vast distances in space.

Nevertheless, astronomers today still find the idea of a celestial sphere very useful. For example, they use it as a means of pinpointing the positions of the stars.

CELESTIAL NORTH AND SOUTH

The celestial sphere represents the skies all around Earth. We divide Earth in two by the Equator, into the Northern and Southern Hemispheres. In a similar way, we divide the celestial sphere in two by a celestial equator, into the northern celestial hemisphere (above Earth's Northern Hemisphere) and the southern celestial hemisphere (above Earth's Southern Hemisphere).

PROFILING—THE STARS

The stars we see in the night sky are distant suns. They are great balls of white-hot gas that pour enormous energy into space as light and other radiation. The main gases in stars are hydrogen and helium. The stars also contain small amounts of many other chemical elements, such as sodium and carbon.

LIGHT-YEARS AWAY

The stars are so far away that it takes years for their light to reach us. Astronomers say that the stars are located light-years away. A "light-year" is the distance light travels in a year, about 6 trillion (million million) miles (10 trillion km).

The closest bright star we can see in the sky (Alpha Centauri) lies just over four light-years away. Others lie hundreds and even thousands of light-years away.

BRIGHTNESS AND COLOR

We can see that the stars differ in brightness. The brightness of a star is called its magnitude. We say that the brightest stars we can see are first magnitude; the dimmest ones are sixth magnitude.

Stars have different colors too. The color indicates a star's temperature. Yellowish stars like the Sun have a temperature at the surface of about 10,000 degrees Fahrenheit (6,000 degrees Celsius). Red stars are much cooler, with only half this temperature. Blue-white stars, on the other hand, can be as much as ten times hotter.

WHERE STARS ARE BORN

Scattered among the stars there are great clouds of gas and dust, which we call nebulas (see page 34). It is here that stars are born.

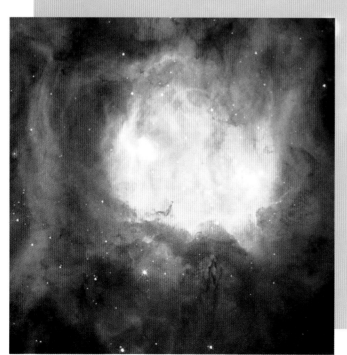

LEFT Stars are born deep inside vast clouds of gas and dust called *nebulas*.

In the clouds, there are pockets of cold gas that are particularly dense (heavy). The clouds start to collapse under gravity—the "pull" of their particles. Over time, the collapsing clouds turn into swirling balls, with a hot mass at their centers. Eventually, the hot mass starts producing nuclear energy and turns into a new star.

LIFE AND DEATH

The new star may then shine steadily for billions of years. The Sun, for example, has been shining for about 5 billion years, and should carry on shining for a similar time.

But eventually all stars have to die. What happens to them depends on how big they are. A small star like the Sun swells up to become a red giant. Then it puffs off a lot of gas and shrinks into a tiny but very dense white dwarf.

PROFILING—THE STARS **027**

LEFT Stars are being born inside these eerie columns within the Eagle Nebula. This spectacular Hubble Space Telescope image has been called the "pillars of creation."

Much bigger stars swell up to supergiant size, then blow themselves to pieces in an explosion called a supernova. What is left after the explosion may become a tiny body called a neutron star, or it may shrink down to almost nothing and disappear in a black hole. A black hole is a region of space with such enormous gravity that it will swallow everything nearby, even light rays.

THE NUCLEAR FURNACE

Stars produce their enormous energy by means of nuclear reactions. These are processes in which atoms, the tiniest particles of matter, take part. In most stars, atoms of hydrogen fuse, or join together, to make helium. This occurs in the center of stars, where pressures are enormous and temperatures reach 27 million degrees Fahrenheit (15 million degrees Celsius) or more.

ABOVE This young star is superhot, with a surface temperature ten times that of the Sun. It is ejecting gas into space at 100,000 miles per hour (160,000 km per hour).

ABOVE A star like the Sun puffs off clouds of gas into space as it dies, forming an object like this, called a *planetary nebula*.

THE SKIES MONTH BY MONTH

The night sky is constantly changing. Every night the constellations seem to move around the sky. And every month new constellations come into view, while others disappear. The same changes take place every year, and at the same times. So we can draw up maps that show which stars are on view at any time of the year.

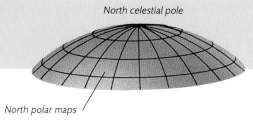

North celestial pole

North polar maps

THE SPINNING HEAVENS

The stars that we see move through the night sky don't really move. They only seem to move because Earth is spinning around in space. Earth spins round from west to east, and so the stars seem to move through the sky in the opposite direction—from east to west. They rise over the horizon in the east and, hours later, set below the horizon in the west, just like the Sun does.

Because the stars move through the night sky, they are changing their positions all the time. So star maps are drawn to show the stars that you can see in the sky in a certain direction at a certain time. The reader should be aware that planets may also appear in some constellations as bright "stars," but these are not shown on the maps.

The star maps that follow show views of the sky, looking north or south at about 10:30 P.M. The maps show skies as they are around the middle of each month.

MONTHLY MOVEMENTS

Which constellations you see in the sky depends not only on the time of night, but also on the time of year. Earth circles around the Sun once a year. This means that, as time goes by, the night side of Earth faces a different part of space, and so you look out at different constellations.

You can trace how the constellations change over the year in the 12 monthly maps. After 12 months, the same constellations come around again.

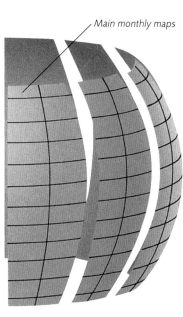

Main monthly maps

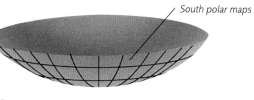

South polar maps

South celestial pole

RIGHT We can split up the celestial sphere into 12 monthly maps, which form the main maps on the pages that follow. The domed maps on the map pages show parts of the northern and southern polar regions of the celestial sphere.

LOOKING AT THE MAPS

The maps are designed so that they can be used by observers in both the Northern and Southern hemispheres. Northern observers need to hold the maps up one way, and southern observers the other. Follow the instructions on the map pages.

MAIN MAPS: Observers working in each hemisphere will see the constellations shown on the main maps, but upside-down from one another. This is because they are looking at them in opposite directions—northern observers are looking south, southern observers are looking north.

POLAR MAPS: The domed maps show broader sky views in front of observers when they turn around to look toward the north (in the Northern Hemisphere) and south (in the Southern). Northern observers will see quite different constellations from southern observers. They will see some constellations that southern observers will never see. And, of course, southern observers will see some constellations that northern observers will never see.

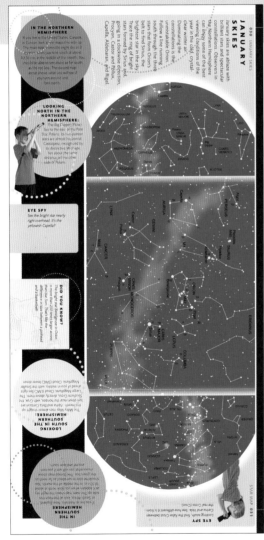

ABOVE If you live in the Northern Hemisphere, look at the main map and the domed map at the top this way up.

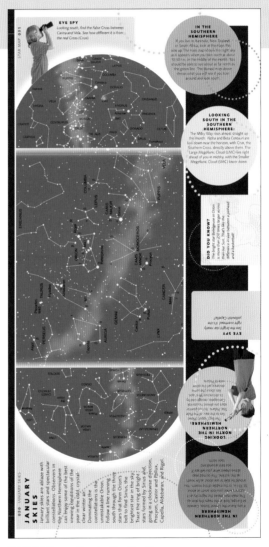

ABOVE If you live in the Southern Hemisphere, look at the main map and the domed map at the top this way up.

JANUARY SKIES

January skies are ablaze with brilliant stars and spectacular constellations. Observers in the Northern Hemisphere can enjoy some of the best viewing conditions of the year in the cold, crystal-clear winter air. Dominating the constellations is the unmistakable Orion. Follow a line running south through the three stars that form Orion's belt to find Sirius, the brightest star in the sky. Trace the ring of bright stars formed by Sirius and, going in a clockwise direction, Procyon, Castor and Pollux, Capella, Aldebaran, and Rigel.

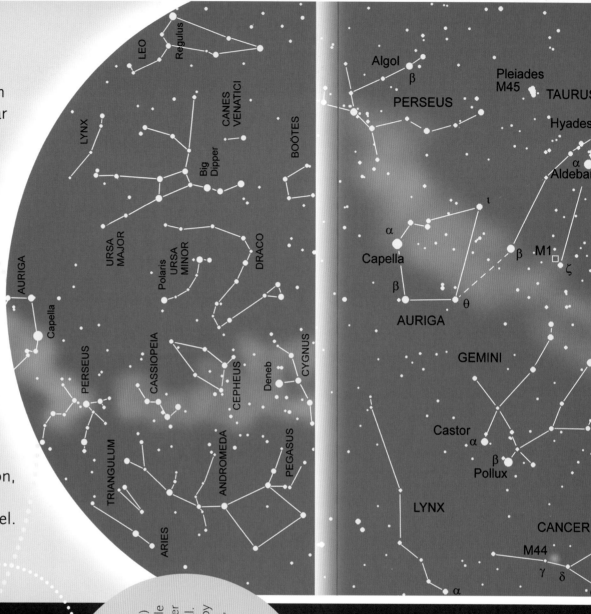

IN THE NORTHERN HEMISPHERE

If you live in the United States, Canada, or Europe, look at the maps this side up. The main map shows the night sky as it appears when you look south at about 10:30 P.M. in the middle of the month. You should be able to see about as far south as the red line. The domed map above shows what you will see if you turn around and look north.

LOOKING NORTH IN THE NORTHERN HEMISPHERE:

The Big Dipper (Plow) lies to the east of the Pole Star, Polaris. Its two pointer stars are almost horizontal. Cassiopeia, recognized by its distinctive W-shape, lies about the same distance on the other side of Polaris.

EYE SPY

See the bright star nearly right overhead? It's the yellowish Capella.

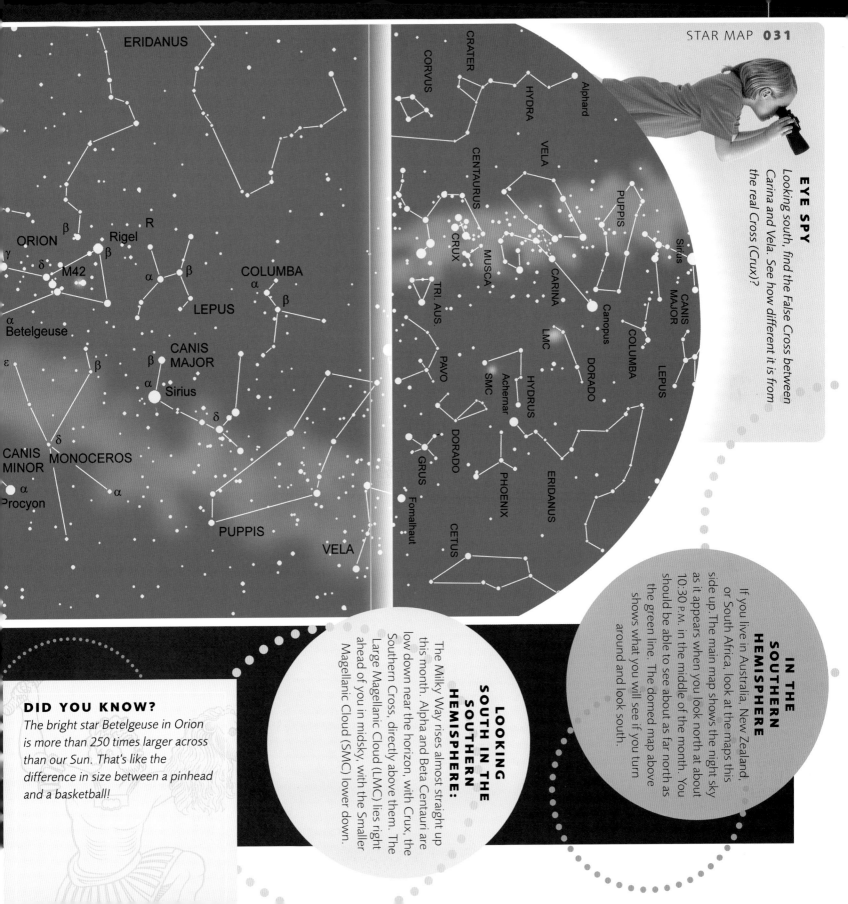

JANUARY
FROM AURIGA TO ORION

AURIGA, THE CHARIOTEER

This northern constellation sits on the Milky Way and is easy to recognize because it is shaped like a kite. Its brightest star is Capella, the sixth brightest star in the whole sky.

In mythology, Auriga represents a charioteer, the driver of a four-horse chariot. The figure has a goat draped over his left shoulder, which is why Capella is also known as the Goat Star. The trio of fainter stars nearby represents the goat's young and are hence called the Kids.

BELOW The brilliant star Capella outshines all the other stars in Auriga. It lies high overhead in winter in the Northern Hemisphere.

CANIS MAJOR, THE GREAT DOG

Canis Major is easy to find because it boasts the brightest of all the stars, Sirius, also called the Dog Star.

The word Sirius means "scorching." The Greeks gave it this name because they believed it brought the heat of summer, as it rises just before the Sun at dawn in July and August.

Sirius is a binary, or double, star, which has another star circling round it, nicknamed the Pup. The Pup is small but dense—just a pinhead of its matter would weigh as much as an elephant! It is a kind of star known as a white dwarf.

CONSTELLATIONS 033

CASSIOPEIA

This northern constellation is one of the easiest to recognize. Its five brightest stars form a conspicuous "W" shape. The constellation represents a vain queen, who was the wife of King Cepheus and mother of Andromeda (represented by other constellations nearby). Cassiopeia boasted that she was the most beautiful of all women, which upset the lovely sea nymphs, the Nereids. They called on the sea god Poseidon (Neptune) to punish her. To find out what happened next, see Andromeda (page 78).

LEPUS, THE HARE

This small constellation lies beneath the feet of the mighty hunter Orion. It is being chased across the sky by the Great Dog, Canis Major. One of the most interesting stars in Lepus is R. Vivid red in color, it is also called the Crimson Star. It is a variable star meaning it varies widely in brightness. At its brightest, it can easily be seen through binoculars.

ORION

This magnificent constellation can easily be seen from both the Northern and Southern Hemispheres. What's more, it really looks like the figure it is supposed to represent: a mighty hunter, with a club in his right hand, ready to land a blow. At his heels in the sky are his dogs, Canis Major and Canis Minor.

Orion's two brightest stars, Betelgeuse and Rigel, are both supergiants. They are easy to tell apart because Betelgeuse is noticeably red, whereas Rigel is brilliant white.

Another highlight of the constellation is the misty patch beneath the three bright stars that mark Orion's belt. This is the Orion Nebula. Turn the page over to find out more about this glorious object and about nebulas in general.

AROUND THE POLE

Most observers in northern North America and northern Europe can see Cassiopeia every night of the year, in different positions. It circles around the Pole Star, making it circumpolar. The constellations Cepheus, Draco, and Ursa Minor and the Big Dipper (Plow) stars are circumpolar too.

NEBULAS— CLOUDS AMONG THE STARS

Look beneath the diagonal made up of three stars in the middle of the constellation Orion, and you'll see a bright misty patch. Through binoculars or a small telescope, it looks like a great glowing cloud. And that's just what it is: a huge cloud of glowing gas, which astronomers call a nebula. The Orion Nebula is one of thousands of nebulas we can see in the sky. But only a long-exposure photograph through a larger telescope will reveal its true size and spectacular colors.

STARS AND SPACE

We tend to think that our Universe is made up of stars traveling through empty space. But the space between the stars—interstellar space—isn't completely empty. It contains tiny traces of gas and dust.

Here and there, wisps of gas and specks of dust may gather together to form thicker, denser clouds, or nebulas. These "dense" clouds are still billions of times thinner than the clouds in the Earth's atmosphere.

BRIGHT NEBULAS

We see many nebulas in the sky when they shine brightly. Sometimes they shine because they reflect the light from nearby stars back toward us. We call these reflection nebulas.

Other nebulas actually shine by themselves. Their gas particles pick up extra energy from the rays given off by hot stars embedded within them. They then emit this extra energy as light. These are called emission nebulas. The Orion Nebula is mainly an emission nebula.

DARK NEBULAS

Many interstellar clouds of gas and dust do not shine at all. We call them dark nebulas. We can see them only when they blot out the light from stars situated behind them. They often look like dark passages or holes in space.

One of the best-known dark nebulas is also found in Orion. It resembles the head and mane of a horse, and hence is called the Horsehead Nebula. In the far Southern Hemisphere there is a dark nebula in the constellation of Crux, the Southern Cross. This is known as the Coal Sack.

LIFE AND DEATH

The cool, dark nebulas in space play an important part in the life of the Universe—they are the places in which stars are born (see page 26).

Some kinds of nebulas also form when stars die. When small stars like the Sun die, they puff off rings of gas to form what are called planetary nebulas. They are called this because through small telescopes they often look like round disks, just like the planets do.

When big stars die, they blast themselves apart in great explosions called supernovas and send out vast clouds of gas and dust into space. Nebulas like this are called supernova remnants.

NEBULAS 035

THE M NUMBERS

The French astronomer Charles Messier (1730–1817) spent his life searching for comets—he was even nicknamed "the comet ferret." But he kept confusing them with misty-looking nebulas and star clusters. So in 1781 he published a list of these objects so that other astronomers wouldn't get confused. Astronomers today still often refer to nebulas and clusters by an M number—the number of the object in Messier's list. The Orion Nebula is M42.

The glorious Orion Nebula. This bright, billowing cloud of gas is lit up by the many hot stars it contains.

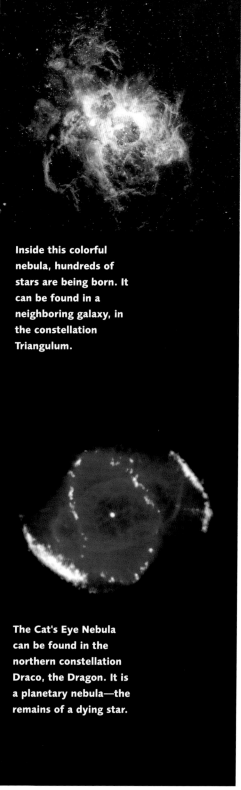

Inside this colorful nebula, hundreds of stars are being born. It can be found in a neighboring galaxy, in the constellation Triangulum.

The Cat's Eye Nebula can be found in the northern constellation Draco, the Dragon. It is a planetary nebula—the remains of a dying star.

FEBRUARY SKIES

February skies are nearly as spectacular as January's, though some duller constellations (like Cancer and Hydra) are moving in from the east. The hexagon, or six-sided figure, made by six bright stars—Sirius, Rigel, Aldebaran, Capella, Pollux, and Procyon—is still visible. It is called the "winter hexagon" by Northern Hemisphere observers. But it also features in summer skies in the Southern Hemisphere, where the seasons are reversed. On a smaller scale, Sirius, Betelgeuse, and Procyon form a "winter triangle" across the Milky Way.

IN THE NORTHERN HEMISPHERE

If you live in the United States, Canada, or Europe, look at the maps this side up. The main map shows the night sky as it appears when you look south at about 10:30 P.M. in the middle of the month. You should be able to see about as far south as the red line. The domed map above shows what you will see if you turn around and look north.

LOOKING NORTH IN THE NORTHERN HEMISPHERE

Gemini's twin stars Castor and Pollux appear high overhead this month, almost directly above Polaris. The Big Dipper has risen higher and Cassiopeia has sunk lower than they were last month. The brilliant skies west of the Milky Way are slipping away, being replaced by duller skies from the east.

EYE SPY
Looking south, find twin stars Castor and Pollux high up. Which do you think is brighter?

STAR MAP 037

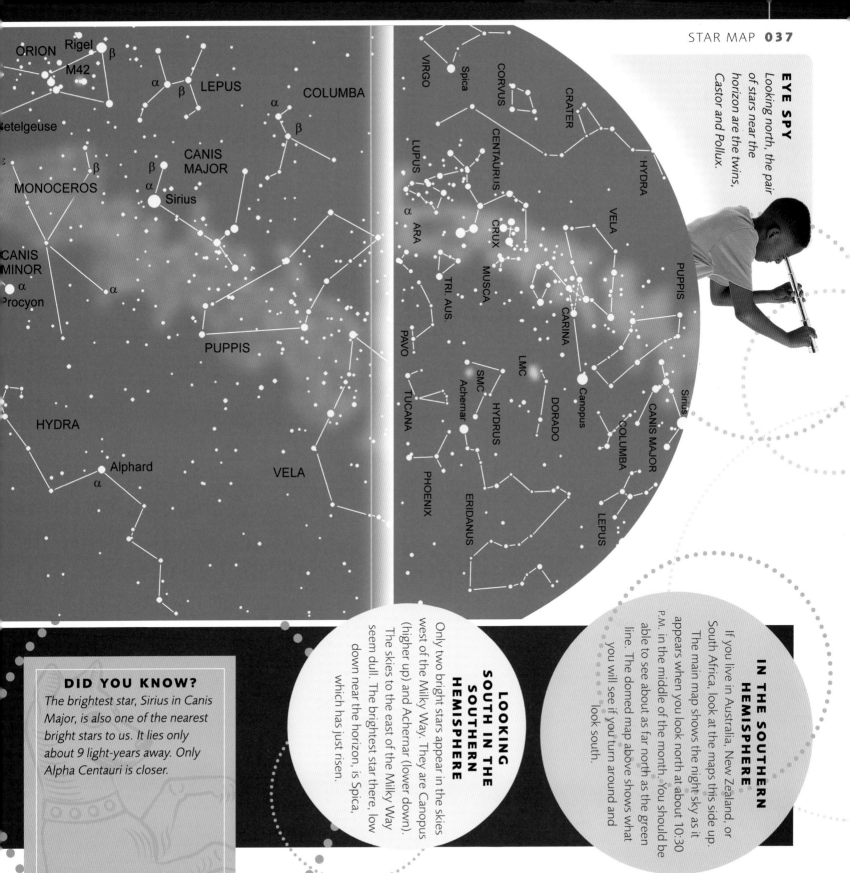

EYE SPY
Looking north, the pair of stars near the horizon are the twins, Castor and Pollux.

IN THE SOUTHERN HEMISPHERE
If you live in Australia, New Zealand, or South Africa, look at the maps this side up. The main map shows the night sky as it appears when you look north at about 10:30 P.M. in the middle of the month. You should be able to see about as far north as the green line. The domed map above shows what you will see if you turn around and look south.

LOOKING SOUTH IN THE SOUTHERN HEMISPHERE
Only two bright stars appear in the skies west of the Milky Way. They are Canopus (higher up) and Achernar (lower down). The skies to the east of the Milky Way seem dull. The brightest star there, low down near the horizon, is Spica, which has just risen.

DID YOU KNOW?
The brightest star, Sirius in Canis Major, is also one of the nearest bright stars to us. It lies only about 9 light-years away. Only Alpha Centauri is closer.

FEBRUARY
FROM CANCER TO GEMINI

CANCER, THE CRAB

Cancer is a faint constellation, but you can find it easily. It lies between two other constellations that are easy to recognize—Leo and Gemini. Like them, Cancer is a constellation of the zodiac (pages 54–55).

The constellation represents the crab that Zeus's wife Hera sent to attack the hero Hercules, whom she hated. But Hercules stamped on the crab and killed it.

An interesting object in this constellation is an open star cluster near the stars Delta (δ) and Gamma (γ). You can just see it with the naked eye, but it looks better through binoculars. Identified as M44, it is called Praesepe, or the Beehive, because it looks like bees buzzing around a hive.

CANIS MINOR, THE LITTLE DOG

This is one of the smallest constellations, dominated by its brightest star, Procyon. This star's name means "before the dog," referring to the fact that it rises in the sky before the Dog Star, Sirius, in Canis Major. Canis Minor and Major represent the two dogs that Orion took hunting. Procyon lies only about 11 light-years away, which makes it the third closest bright star in the sky, after Alpha Centauri and Sirius.

MONOCEROS, THE UNICORN

This isn't a brilliant constellation, but it can easily be found between the two "Dogs," Canis Major and Minor. It represents the fabled beast, a horse with a single horn. It is one of the later constellations (defined in the 1600s), not recognized by ancient astronomers. Monoceros has no really bright stars, but lies on the Milky Way and is worth scanning in binoculars. Two fine nebulas can be seen north of Epsilon (ε)—the Rosette and the Cone.

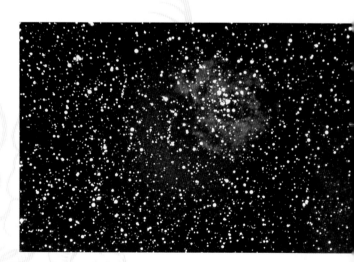

ABOVE Monoceros is flooded with gas and dust. One bright gas cloud is the petal-like Rosette Nebula.

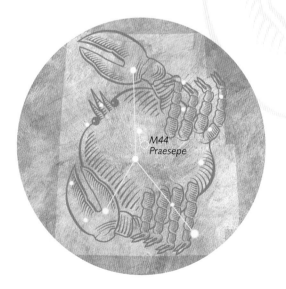

M44 Praesepe

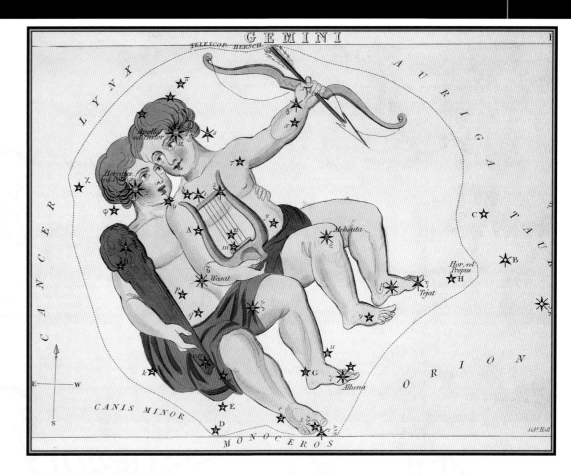

PUPPIS, THE POOP

Northern astronomers have perhaps the best view of this southern constellation this month. Just south of Canis Major, it straddles the Milky Way. It looks glorious through binoculars, which reveal rich star fields and many star clusters. Puppis was once part of the much larger constellation Argo Navis, named for the ship on which Jason and the Argonauts sailed on their quest for the Golden Fleece. The constellation was split into three—Puppis (Poop), Carina (Keel), and Vela (Sails)—in the 1700s.

GEMINI, THE TWINS

Gemini is one of the trio of constellations that dominate winter skies in the Northern Hemisphere and summer skies in the Southern. The others are Orion and Taurus. Gemini is a constellation of the zodiac (pages 54–55).

The constellation's most prominent stars are the bright pair Castor and Pollux. They represent the twin sons of Zeus, king of the gods, and Leda, queen of Sparta. They hatched out of eggs that Leda laid. Pollux is the brighter of the two, but Castor is more interesting. It is a multiple star system in which six stars revolve around one another.

ABOVE In mythology, Castor and Pollux were identical twins who did everything together and were inseparable. When they grew up, Castor became a great horseman, and Pollux became a champion boxer.

MARCH SKIES

In the Northern Hemisphere, days are lengthening rapidly, and spring begins on or around March 20. In the Southern Hemisphere on the other hand, days are shortening, and March 20 signals the beginning of fall. March 20 is the date when the Sun appears to cross over the equator from the Southern to the Northern Hemisphere. At this time, the lengths of day and night are the same throughout the world (12 hours). It is called the spring, or vernal, equinox. (*Equinox* means "equal night.") In the sky, it is the appearance of Leo in the skies that heralds spring (or fall).

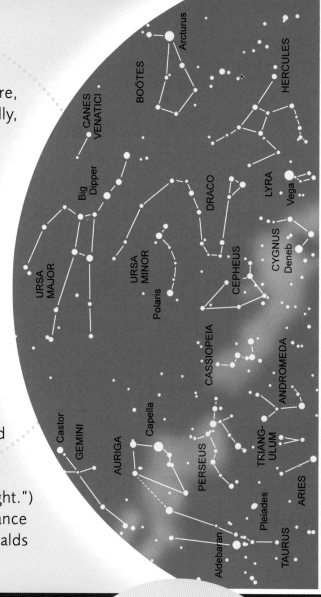

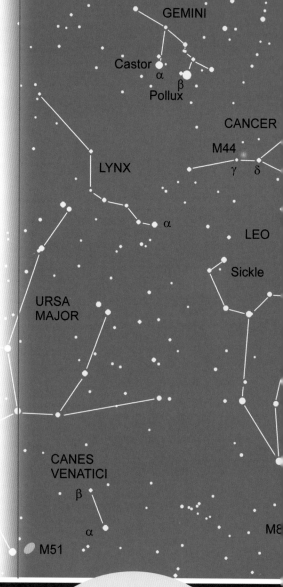

EYE SPY Looking south, notice that Sirius and Rigel are low down in the west.

IN THE NORTHERN HEMISPHERE If you live in the United States, Canada, or Europe, look at the maps this side up. The main map shows the night sky as it appears when you look south at about 10:30 P.M. in the middle of the month. You should be able to see about as far south as the red line. The domed map above shows what you will see if you turn around and look north.

LOOKING NORTH IN THE NORTHERN HEMISPHERE March is a good month to spot Cepheus, because it lies straight ahead, directly beneath Polaris and to the right of Cassiopeia. In the west, Capella (highest) and Aldebaran are descending. Note the difference in color between them.

STAR MAP 041

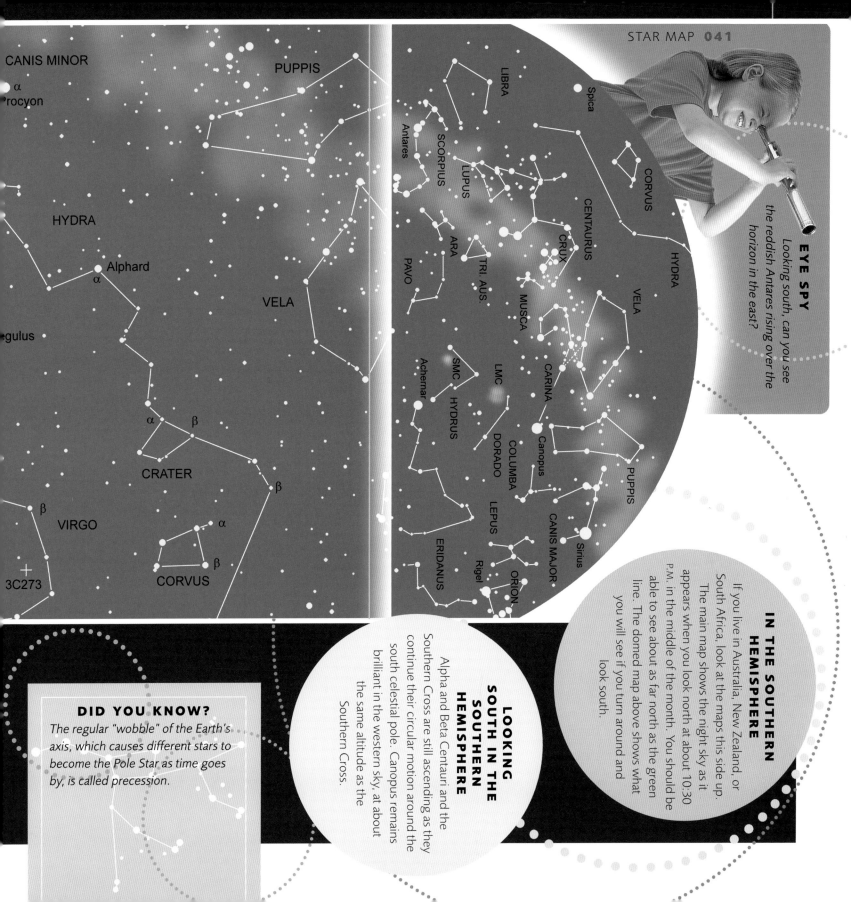

EYE SPY
Looking south, can you see the reddish Antares rising over the horizon in the east?

IN THE SOUTHERN HEMISPHERE
If you live in Australia, New Zealand, or South Africa, look at the maps this side up. The main map shows the night sky as it appears when you look north at about 10:30 P.M. in the middle of the month. You should be able to see about as far north as the green line. The domed map above shows what you will see if you turn around and look south.

LOOKING SOUTH IN THE SOUTHERN HEMISPHERE
Alpha and Beta Centauri and the Southern Cross are still ascending as they continue their circular motion around the south celestial pole. Canopus remains brilliant in the western sky, at about the same altitude as the Southern Cross.

DID YOU KNOW?
The regular "wobble" of the Earth's axis, which causes different stars to become the Pole Star as time goes by, is called precession.

MARCH
FROM CARINA TO VELA

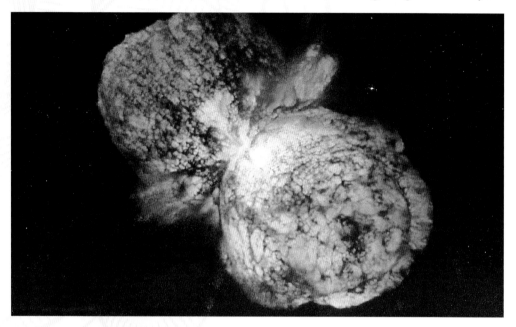

BELOW The massive star Eta Carinae often flares up and ejects vast quantities of glowing matter into space.

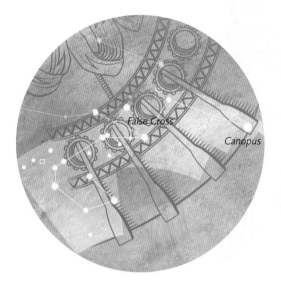

CARINA, THE KEEL

Carina is a fine constellation of the far southern skies, though unfortunately beyond the view of most northern astronomers. It is one of the three constellations created out of the original large constellation Argo Navis (see Puppis, page 39).

Carina's dominant star is Canopus, the second brightest star in the sky, after Sirius. It is a white supergiant star, nearly 15,000 times brighter than our Sun. Two of Carina's stars make a cross shape with two stars in neighboring Vela. Together they form the False Cross, which can sometimes be mistaken for the true Southern Cross, Crux.

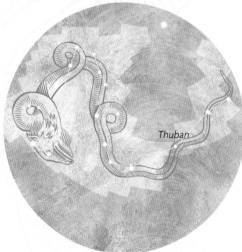

DRACO, THE DRAGON

This sprawling northern constellation winds itself around the Pole Star, Polaris. It represents the dragon that guarded the fabulous golden apples in the garden of the beautiful Hesperides. These were the daughters of Atlas, the god who carried the heavens on his shoulders. The hero Hercules slew the dragon and stole the golden apples on one of his Labors (see page 58).

About 4,000 years ago, Draco's brightest star Thuban used to be the Pole Star—the star directly above the Earth's North Pole. But since then, the regular wobbling of the Earth's axis has brought Polaris into position as Pole Star.

CONSTELLATIONS

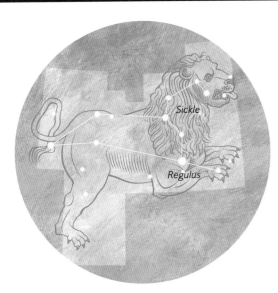

ABOVE The original constellation Argo Navis, depicted in 1801 star atlas *Uranographia*, compiled by German astronomer Johann Bode.

LEO, THE LION

Leo is one of the few constellations that look like the figure they are supposed to represent, in this case a crouching lion. It represents the Nemean lion that Hercules had to fight on the first of his Labors (see page 58). He couldn't kill it with a spear or arrows because they just bounced off its skin. So Hercules had to strangle it with his bare hands.

The easiest part of Leo to recognize is the curve of stars forming the lion's head and chest. Because of its shape, this constellation is also called "the Sickle." The brightest star, Regulus ("little king"), is also sometimes called Cor Leonis, meaning "heart of the lion." Leo is a constellation of the zodiac (see pages 54–55).

URSA MAJOR, THE GREAT BEAR

Ursa Major is the larger of the two bears that grace far northern skies. The other is Ursa Minor. Observers in the Southern Hemisphere may snatch a glimpse of this constellation near the horizon this month.

Ursa Major represents the nymph Callisto, who was wooed by Zeus, the king of the gods. Zeus's wife Hera didn't think much of this and turned Callisto into a bear.

The seven brightest stars in Ursa Major make up one of the best-known star patterns in the sky. The pattern resembles the shape of an old-fashioned milk ladle and is called the Big Dipper. (In Britain, the pattern is known as the Plow, because it also resembles the handle and blade of a hand-plow.) The Big Dipper is a useful signpost to other stars (see pages 44–45).

VELA, THE SAILS

Vela was once part of the large constellation Argo Navis (see Puppis, page 39). Two of its stars form the False Cross with two stars in Carina (see above). The constellation is set in a brilliant part of the Milky Way and is full of star clusters and nebulas.

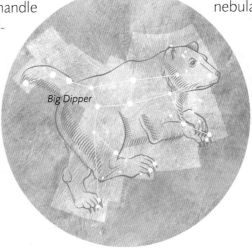

KEY CONSTELLATIONS—SIGNPOSTS TO THE STARS

If you live in northern parts of the world, the Big Dipper (Plow) is an old friend. You should be able to see its ladle shape on almost every night of the year. From many northern latitudes it is circumpolar (which means that it never goes below the horizon), and circles round the Pole Star, Polaris. The Big Dipper is unmistakable—and very useful. Astronomers often describe it as a "signpost," because you can use it to point to other stars and constellations. Other constellations act as useful signposts too, particularly Orion.

THE POINTERS

Because they are so prominent, all seven stars in the Big Dipper pattern have names. Arab astronomers gave them the names we use today. The two stars at the ladle end of the pattern are named Merak and Dubhe. These stars are called the Pointers because if you follow an imaginary line through them, you will come to the Pole Star, Polaris. When you have found Polaris, you will be able to trace the shape of Ursa Minor, or the Little Dipper.

TO THE SWAN AND THE LION

Going back to the Big Dipper, follow a line first through the two other ladle stars, Phekda and Megrez. If you continue the line through the brightest star in the ladle end of the Little Dipper, you will eventually come to a much brighter star. This is Deneb, at the tail end of the swan, Cygnus, also called the Northern Cross.

If you follow a similar line going in the opposite direction from Phekda and Megrez, you'll reach another bright star. It is Regulus, in the constellation of the lion, Leo.

Both Dippers act as signposts to other stars and constellations too, as the Dippers map shows.

NORTH AND SOUTH

Unfortunately, the Dippers act as useful signposts only in the Northern Hemisphere. They are too far north to be of any use to southern observers. On the other hand, southern observers have signpost stars such as Alpha and Beta Centauri that northern observers can't see.

FOLLOW THAT STAR

Fortunately, there is an absolutely invaluable signpost that can be seen equally well in both northern and southern skies. It is the unmistakable figure of the mighty hunter, Orion, which straddles the celestial equator.

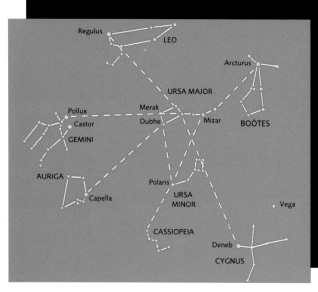

KEY CONSTELLATIONS **045**

THE SOUTHERN POINTERS

Alpha and Beta Centauri are brilliant pointers in far southern skies. Follow a line through them, and you'll quickly come to the most famous southern constellation, Crux, the Southern Cross. Without Alpha and Beta Centauri to guide you, you might be tempted to mistake the False Cross—not far away—for the true Southern Cross. The False Cross is shaped like a cross and made by a pair of stars from Vela and a pair from Carina.

LEFT In the far southern heavens, the twin bright stars Alpha and Beta Centauri (to the right of the picture) point directly to Crux, the Southern Cross.

Follow a line through the three stars in Orion's belt toward the south, and you'll find the brilliant Sirius, the Dog Star. Following a line northward through the three stars takes you to Aldebaran, the eye of the bull. Continue the line and you'll reach the Pleiades star cluster.

Other alignments using the stars in Orion will take you to all surrounding bright stars and constellations, as the Orion map shows.

BELOW Follow a line through Orion's Belt and orange star Aldebaran, and you'll come to this fine cluster, the unmistakable Pleiades.

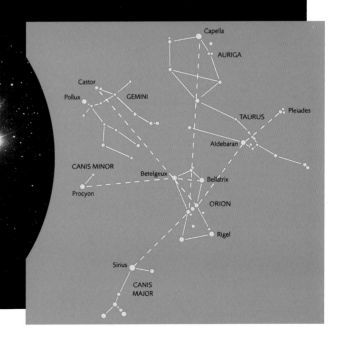

APRIL SKIES

In April, there is still plenty of interest in far northern and far southern skies, but the skies in between are relatively bare, apart from Leo. Two faint constellations occupy huge swathes of sky—Hydra and Virgo, which are the two biggest of the constellations. Virgo's only bright star, Spica, forms a worthy "spring (or fall) triangle" with Leo's Regulus and Boötes's Arcturus. In the Northern Hemisphere, the Milky Way is too close to the horizon for decent viewing. In the Southern Hemisphere, however, it sits in midsky and is well placed for observation.

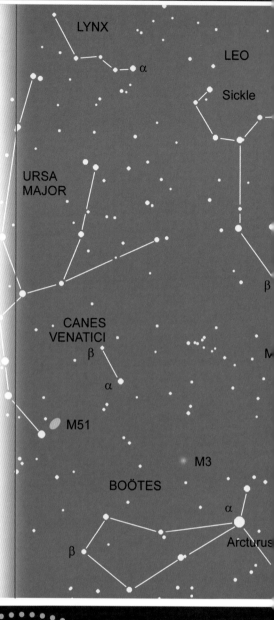

IN THE NORTHERN HEMISPHERE

If you live in the United States, Canada, or Europe, look at the maps this side up. The main map shows the night sky as it appears when you look south at about 10:30 P.M. in the middle of the month. You should be able to see about as far south as the red line. The domed map above shows what you will see if you turn around and look north.

LOOKING NORTH IN THE NORTHERN HEMISPHERE

Cassiopeia is reaching its lowest point in the sky, almost directly below Polaris. The Big Dipper (Plow), on the other hand, has climbed to its highest point, with the handle roughly parallel with the horizon. Deneb and Vega have risen in the east and are climbing.

EYE SPY

Looking north, can you spot that Nizar, the second star in the handle of the Big Dipper is a double with Alcor?

STAR MAP 047

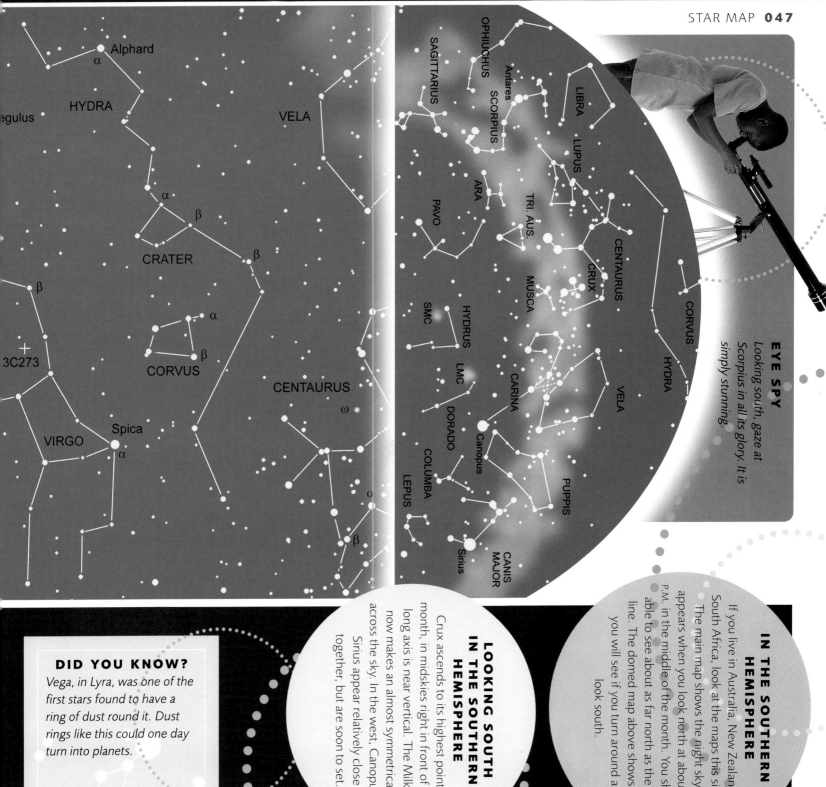

EYE SPY
Looking south, gaze at Scorpius in all its glory. It is simply stunning.

IN THE SOUTHERN HEMISPHERE
If you live in Australia, New Zealand, or South Africa, look at the maps this side up. The main map shows the night sky as it appears when you look north at about 10:30 P.M. in the middle of the month. You should be able to see about as far north as the green line. The domed map above shows what you will see if you turn around and look south.

LOOKING SOUTH IN THE SOUTHERN HEMISPHERE
Crux ascends to its highest point this month, in midskies right in front of you. Its long axis is near vertical. The Milky Way now makes an almost symmetrical arch across the sky. In the west, Canopus and Sirius appear relatively close together, but are soon to set.

DID YOU KNOW?
Vega, in Lyra, was one of the first stars found to have a ring of dust round it. Dust rings like this could one day turn into planets.

APRIL
FROM CANES VENATICI TO VIRGO

CANES VENATICI, THE HUNTING DOGS

This small constellation can be found under the curved handle of the Big Dipper (Plow). It represents two dogs on a leash held by the charioteer Boötes. They are snapping at the feet of the Great Bear, Ursa Major.

Canes Venatici has two delights, both visible through binoculars. The first is a fine globular cluster, M3. Look for it roughly halfway along a line between Alpha and the bright star Arcturus in Boötes. The other delight is M51, the Whirlpool Galaxy. It lies just south of the first star in the handle of the Big Dipper.

CORVUS, THE CROW

Corvus perches on the back of the Water Snake, Hydra. The constellations feature in a story in which the god Apollo asked a crow to get water for him in a cup (the constellation Crater). But the crow spent time eating fruit instead, then blamed the water snake for stopping him from getting the water. So Apollo put the three constellations in the heavens, with the crow never able to quench its thirst because the cup is forever out of its reach.

Corvus is a tiny constellation. Its main pattern is made up of four almost equally bright stars. They make a nice group when viewed through binoculars.

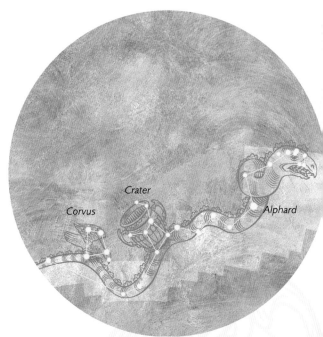

HYDRA, THE WATER SNAKE

The largest constellation of all, Hydra winds itself a quarter of the way around the heavens. It represents the multiheaded serpent, with poisonous breath, that Hercules had to fight on the last of his Labors (page 58). After he had killed it, Hercules dipped his arrows in the Hydra's blood, which made them deadly. The water snake also features in another well-known story, with Corvus and Crater (see above).

Hydra is big but not particularly interesting, passing through a relatively empty region of the heavens. It has only one reasonably bright star, called Alphard, meaning "the Solitary One."

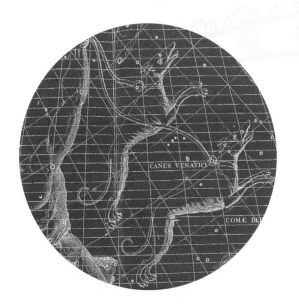

VIRGO, THE VIRGIN

After Hydra, Virgo is the biggest constellation in the sky, but it isn't that easy to find. The best way may be first to find the bright star Arcturus at the foot of the prominent kite-shaped constellation, Boötes, and then to look south. The next bright star you'll spot is Virgo's main star, Spica.

Virgo usually represents Demeter (Ceres), the goddess of agriculture, and is depicted with an ear of wheat in her left hand (marked by Spica).

Astronomically, the outstanding feature of Virgo is its galaxies. They gather together in their thousands to form a huge cluster. A few (like M87) can be seen in small telescopes, but most require larger ones. Another historically important object in the constellation, again visible only in large telescopes, is 3C-273. It was the first quasar to be identified.

INCREDIBLE QUASARS

Quasars are the most incredible bodies. They look like stars in telescopes, but are much more distant than the stars in the sky. They are as bright as hundreds of galaxies put together!

These are some of the thousands of quasars astronomers have found. These pictures were taken by the Hubble Space Telescope. Most quasars lie billions of light-years away.

MAY SKIES

Faint constellations again dominate much of the skies in May. Libra and Serpens have now joined Hydra and Virgo. The bright pair of stars Arcturus and Spica now occupy center stage and make an interesting contrast. Arcturus is a distinctly orange color, whereas Spica is brilliant white. As with all stars, the difference in color reflects a difference in temperature. Arcturus is cool, with a surface temperature of around 7,000 degrees Fahrenheit (4,000 degrees Celsius). Spica is incredibly hot, with a temperature of more than 45,000 degrees Fahrenheit (25,000 degrees Celsius).

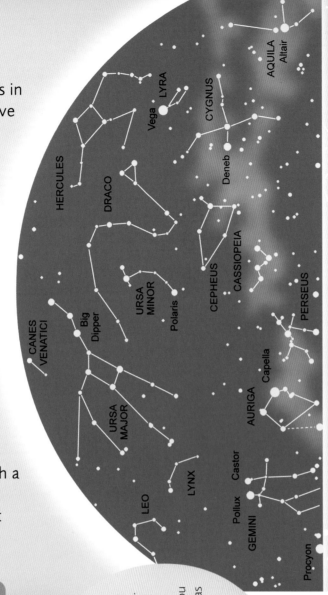

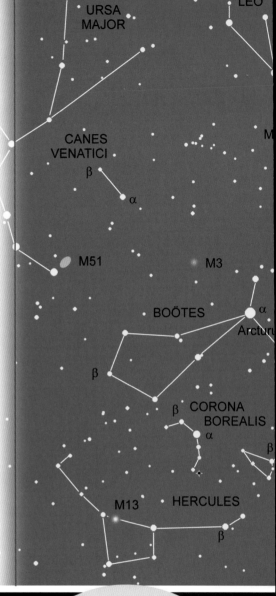

EYE SPY
Looking north, see the arc of bright stars low down in the west.

IN THE NORTHERN HEMISPHERE
If you live in the United States, Canada, or Europe, look at the maps this side up. The main map shows the night sky as it appears when you look south at about 10:30 P.M. in the middle of the month. You should be able to see about as far south as the red line. The domed map above shows what you will see if you turn around and look north.

LOOKING NORTH IN THE NORTHERN HEMISPHERE
Cassiopeia is still low down in the sky and the Big Dipper high up. The distinctive bird shape of Cygnus is well risen in the east, with its brightest star Deneb joining Vega higher up to make a prominent pair. Meanwhile, another bird (Aquila) is just rising over the eastern horizon.

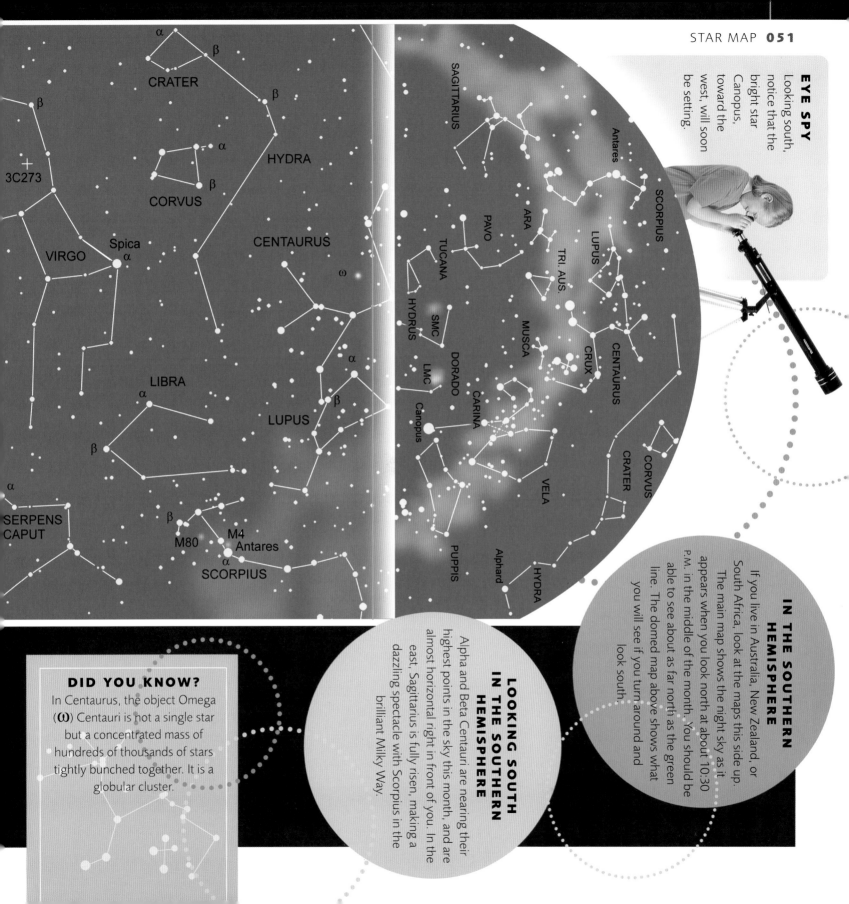

MAY
FROM BOÖTES TO URSA MINOR

BOÖTES, THE HERDSMAN
Boötes is a northern constellation that is easy to identify—it is shaped like a kite. Its leading star is the brilliant Arcturus, which is the brightest star in the northern celestial hemisphere and the fourth brightest star in the whole heavens.

The actual word *Arcturus* means something like "bear keeper," and it is easily found by following the curve of the handle of the Big Dipper (Plow) southward. The handle, of course, represents the tail of the Great Bear, Ursa Major. And Boötes represents a man herding, or chasing, a bear. In mythology, he was Arcas, who ended up chasing his mother Callisto, who had been turned into a bear.

LIBRA, THE SCALES
Libra is quite a faint constellation, but it can be found fairly easily because it lies just to the north of the brilliant Scorpius. It represents the scales of justice held by the figure of Virgo, who was once considered the goddess of justice.

Libra's two brightest stars boast perhaps the most delightful names in astronomy. Alpha is named Zubenelgenubi; Beta is called Zubenelschemali. They are Arabic names, meaning, respectively, "southern claw" and "northern claw." This dates back to the time when the stars were considered to be part of Scorpius.

Libra's main claim to fame is that it is a constellation of the zodiac. Find out more about the zodiac on page 54.

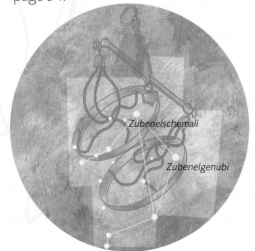

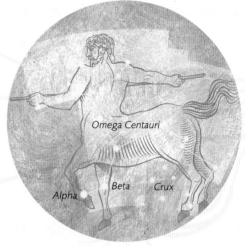

CENTAURUS, THE CENTAUR
The Centaur is one of the most stunning constellations in the heavens, containing a host of spectacular objects. It is a favorite constellation for observers in the Southern Hemisphere. Unfortunately, most observers in the Northern Hemisphere can only glimpse some of its stars, low down on the horizon this month.

Centaurus represents a centaur, a fabulous creature that was half-man, half-horse. Most centaurs were wild creatures, but this centaur was identified with a quieter, learned figure named Chiron, skilled in the arts and medicine.

LEFT The peculiar galaxy Centaurus A, which appears divided in two by a dark dust lane. It is a radio galaxy, that emits powerful radio waves. Astronomers believe that a massive black hole produces its enormous energy.

The two brightest stars in Centaurus, Alpha and Beta, are known as the southern pointers because they point to Crux, the Southern Cross. Other notable objects are the globular cluster Omega (ω) Centauri and Centaurus A, a galaxy that pumps out enormous energy as radio waves.

URSA MINOR, THE LITTLE BEAR

Ursa Minor is the smaller of the two "bears" in the far northern celestial hemisphere—the other is Ursa Major. The pattern of the main stars in Ursa Minor is a smaller, fainter version of the Big Dipper in Ursa Major. So it is called the Little Dipper.

By far the most important star in the Little Dipper is the one at the end of the "handle." Named Polaris, it is also called the North Star and the Pole Star, because it lies almost directly above the Earth's North Pole. Polaris varies slightly in brightness. It is a kind of variable star called a Cepheid.

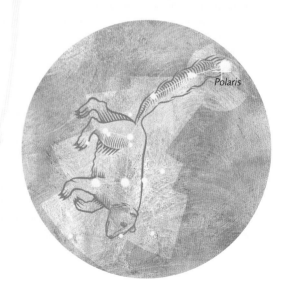

THE ZODIAC—
THE CIRCLE OF ANIMALS

Every year, the Sun appears to travel through the sky against a background of stars. It travels through the same stars—the same constellations—at the same time every year. The Moon and all the planets are always found within these same constellations, too. Ancient astronomers called them the constellations of the zodiac, a word that means something like "the circle of animals." It gained this name because most constellations are represented as animal figures—such as Leo (the Lion), Scorpius (the Scorpion), and Taurus (the Bull).

ABOVE "Buzz" Aldrin on the Moon in 1969. Born on January 20, 1930, "Buzz" seems to be a typical Aquarian, fascinated by science and technology and eager to explore.

BRIGHT AS DAY
Of course, we can't actually see the Sun traveling through the stars, because its dazzling light hides the much fainter stars from view. But if the Sun were only as bright as the Moon, then we would see it traveling through the constellations of the zodiac and passing through a different constellation more or less every month.

Today, astronomers know that the constellations of the zodiac are nothing special. Indeed, many of these constellations (such as Pisces and Capricornus) are faint and contain little interest astronomically.

AN ANCIENT BELIEF
In ancient times, people worshiped the Sun as a god, so they thought the constellations it passes through during the year must be special. They believed that these constellations somehow affect people's characters and the lives they lead.

This belief is known as astrology. It must not be confused with astronomy. Astronomy is the scientific study of the stars and the celestial bodies. Astrology is the study of the stars and planets to find how they might affect human lives and events.

LEFT Leo, the Lion, is one of the easiest constellations of the zodiac to recognize.

STAR SIGNS

About 2,000 years ago, astrologers realized that the Sun passed through about 12 constellations during the year, roughly one every month. So they divided up the year into 12 monthly periods and assigned each one to the constellation the Sun passed through that month. The 12 periods covered by the 12 constellations, or signs, of the zodiac are shown in the table.

Astrologers think that the sign under which you were born is important, as are the positions of the planets in the other signs (zodiac constellations) when you were born. These things are supposed to determine what kind of person you are and what might happen to you.

To find information about you and your life, astrologers prepare a birth chart. This chart shows the star sign under which you were born (the "Ascendant") and the positions of the planets in the other signs (or "houses") at the time of your birth.

Astronomers find no evidence that astrology works. In fact, much is working against it. For example, present-day astrology is still based on the skies as they were 2,000 years ago. Since then, the sky has moved on, and the Sun now passes through the constellations a month earlier than it did in ancient times.

SIGNS OF THE ZODIAC

1st	♈	Aries	March 21–April 19
2nd	♉	Taurus	April 20–May 20
3rd	♊	Gemini	May 21–June 21
4th	♋	Cancer	June 22–July 22
5th	♌	Leo	July 23–August 22
6th	♍	Virgo	August 23–September 22
7th	♎	Libra	September 23–October 22
8th	♏	Scorpius	October 23–November 21
9th	♐	Sagittarius	November 22–December 21
10th	♑	Capricornus	December 22–January 19
11th	♒	Aquarius	January 20–February 18
12th	♓	Pisces	February 19–March 20

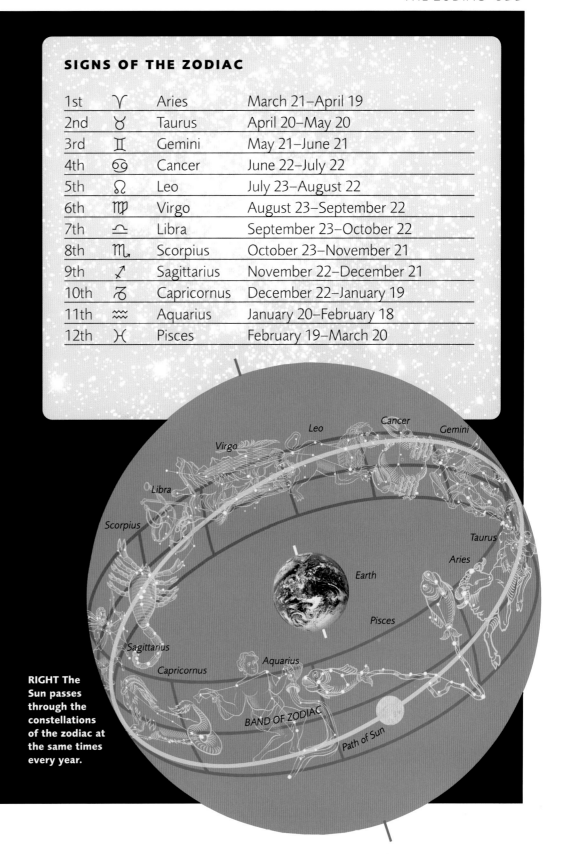

RIGHT The Sun passes through the constellations of the zodiac at the same times every year.

056 JUNE SKIES

JUNE SKIES

In the Northern Hemisphere, the approach of midsummer means warm nights, but lengthening twilight, making stargazing more difficult. Observers in the Southern Hemisphere have no such problems, because down there it is approaching midwinter and skies are dark and clear. During this month and the next, northern observers can catch sight of Scorpius above the southern horizon. But southern observers see this unmistakable constellation in all its glory, along with Sagittarius and the other constellations in the brilliant band of the Milky Way.

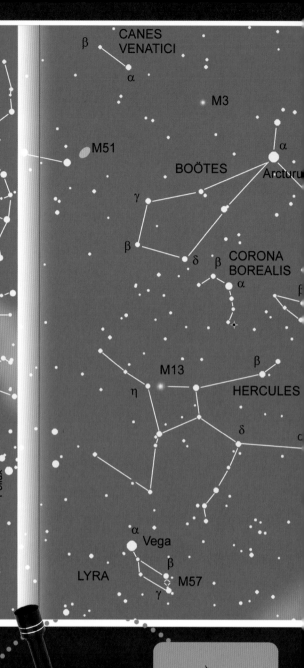

IN THE NORTHERN HEMISPHERE

If you live in the United States, Canada, or Europe, look at the maps this side up. The main map shows the night sky as it appears when you look south at about 10:30 P.M. in the middle of the month. You should be able to see about as far south as the red line. The domed map above shows what you will see if you turn around and look north.

LOOKING NORTH IN THE NORTHERN HEMISPHERE

Ursa Minor, the Little Dipper, is standing on its tail as it reaches its highest point in the sky. The Big Dipper is now descending and Cassiopeia rising. Deneb is at about the same altitude as Polaris and almost directly beneath the brighter Vega.

EYE SPY

Looking north, see how the pointer stars in the Big Dipper almost point to Deneb.

STAR MAP 057

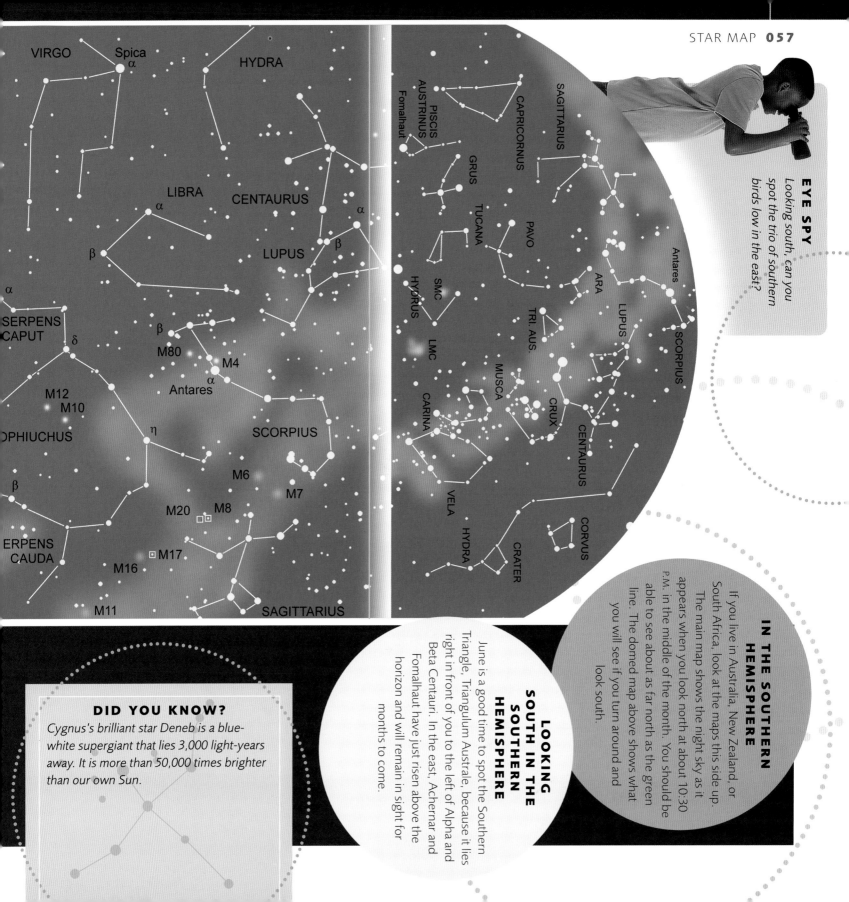

EYE SPY
Looking south, can you spot the trio of southern birds low in the east?

IN THE SOUTHERN HEMISPHERE
If you live in Australia, New Zealand, or South Africa, look at the maps this side up. The main map shows the night sky as it appears when you look north at about 10:30 P.M. in the middle of the month. You should be able to see about as far north as the green line. The domed map above shows what you will see if you turn around and look south.

LOOKING SOUTH IN THE SOUTHERN HEMISPHERE
June is a good time to spot the Southern Triangle, Triangulum Australe, because it lies right in front of you to the left of Alpha and Beta Centauri. In the east, Achernar and Fomalhaut have just risen above the horizon and will remain in sight for months to come.

DID YOU KNOW?
Cygnus's brilliant star Deneb is a blue-white supergiant that lies 3,000 light-years away. It is more than 50,000 times brighter than our own Sun.

JUNE
FROM CEPHEUS TO SCORPIUS

CEPHEUS
Most Northern Hemisphere observers can see this constellation every night, because it lies close to the Pole Star. It represents King Cepheus, the husband of the vain Queen Cassiopeia and father of Andromeda, who was so nearly gobbled up by the sea monster Cetus (page 78).

Cepheus has two interesting stars, both variables. One is Delta (δ), which varies in brightness as regularly as clockwork and gives its name to the class of variables called Cepheids. The other is Mu (μ), which is called the Garnet Star because of its rich red color.

CORONA BOREALIS, THE NORTHERN CROWN
Corona Borealis is a little semicircle of stars wedged between the more brilliant constellations Boötes and Hercules. It represents the jeweled crown of Ariadne, daughter of Minos, king of Crete. She was the half-sister of the terrifying bull-headed creature called the Minotaur. Ariadne married Dionysus (Bacchus), the god of wine, who flung her crown into the heavens, where its jewels turned into stars. The name of its brightest star, Gemma, is the Latin word for "jewel."

HERCULES
The northern constellation Hercules is the fifth largest in the heavens, but has no really outstanding stars. It represents one of the greatest of the ancient Greek heroes, who was a son of Zeus and of a beautiful mortal woman named Alcmene. Zeus's real wife Hera hated Hercules and made him kill his own children.

As penance for his dreadful crime, Hercules was ordered to work for his cousin King Eurystheus, who set him 12 apparently impossible tasks. They became known as the Labors of Hercules. Among these Labors, Hercules strangled a ravaging lion, slew terrifying monsters such as the Hydra, and also captured Cerberus, the ferocious watchdog of the underworld, which had three heads and a coat of poisonous snakes.

An outstanding sight in Hercules is M13, which is one of the brightest globular clusters. It is just visible to the naked eye and is easily seen through binoculars. Look for it just south of the star Zeta (ζ).

LEFT The globular cluster M4, near Antares, in Scorpius.
BELOW Close-up of the stars in M4. The circled ones are white dwarfs.

PAVO, THE PEACOCK

Pavo is one of the "southern birds," which "fly" in circles around the southern celestial pole. Like the others—Tucana (the Toucan), Grus (the Crane), and Phoenix (the Phoenix)—it is a later constellation (defined in the 1500s), not known to ancient astronomers. Alpha (α), also named the Peacock, is by far the brightest of Pavo's stars.

SCORPIUS, THE SCORPION

Scorpius is undoubtedly one of the most spectacular constellations in the sky. And it is one of the few that resembles the figure it is named after, with a curve of bright stars just like the deadly curved tail of a scorpion. You need to be in the Southern Hemisphere to appreciate Scorpius fully. But Northern observers get their best view of it this month.

The constellation represents the scorpion that killed the mighty hunter Orion. It is a constellation of the zodiac (see pages 54–55).

Scorpius's brightest star is Antares. Its name means "rival of Mars." This refers to the fact that it has a noticeably reddish color, like the planet Mars. Most of the constellation lies in one of the brightest regions of the Milky Way, and so looks stunning through binoculars. Use them to look at three beautiful clusters, M4 near Antares, and M6 and M7 just north of the scorpion's tail.

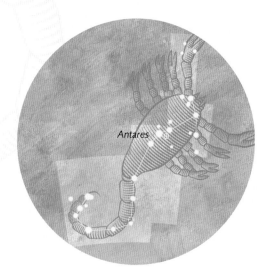

JULY SKIES

In this month and next in the summer skies of the Northern Hemisphere, Cygnus, Lyra, and Aquila sit high overhead. Their three brightest stars—Deneb, Vega, and Altair—declare that summer has arrived. They form the celebrated "summer triangle." In the Southern Hemisphere, the trio appear low in the sky and serve as proof that winter has arrived. The Milky Way provides a feast to the eyes in both hemispheres, although northern observers may be hampered by light summer skies. Nevertheless, these skies continue to offer the best views of Sagittarius and Scorpius in the far south.

IN THE NORTHERN HEMISPHERE

If you live in the United States, Canada, or Europe, look at the maps this side up. The main map shows the night sky as it appears when you look south at about 10:30 P.M. in the middle of the month. You should be able to see about as far south as the red line. The domed map above shows what you will see if you turn around and look north.

LOOKING NORTH IN THE NORTHERN HEMISPHERE

Vega and Deneb are climbing ever higher into the heavens as summer arrives. But in the east the Square of Pegasus is rising, as a reminder that fall isn't so far away. On the horizon, Capella is almost directly beneath Polaris.

EYE SPY

Looking south, can you see an orange-colored star near the horizon? This is Antares.

STAR MAP 061

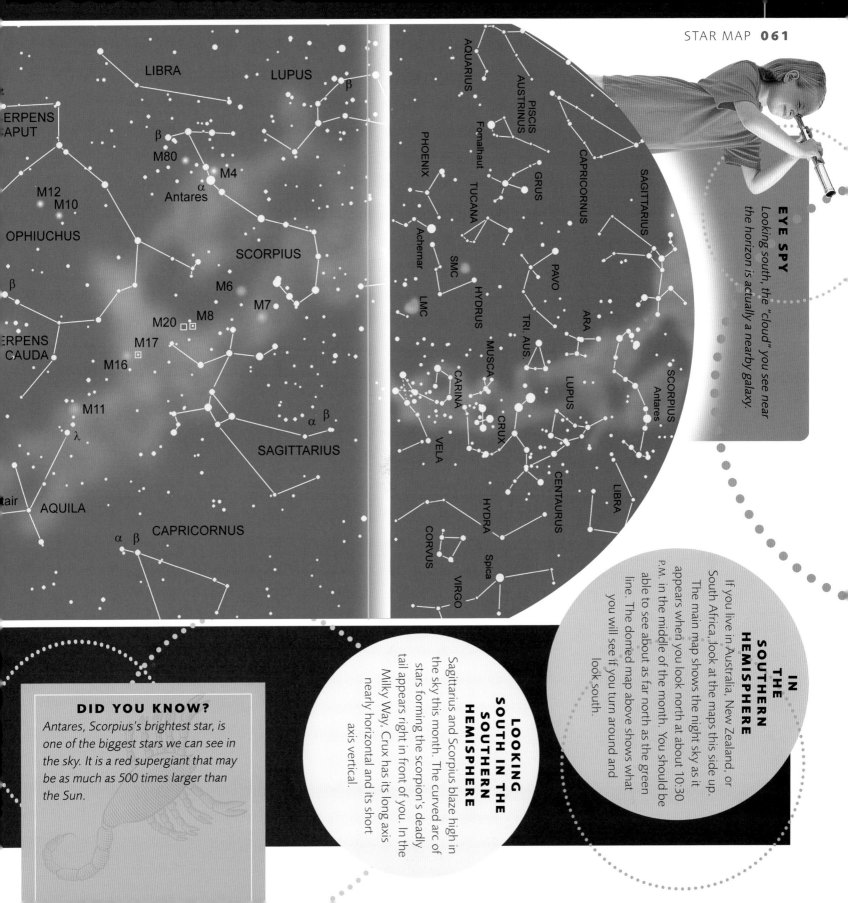

EYE SPY
Looking south, the "cloud" you see near the horizon is actually a nearby galaxy.

IN THE SOUTHERN HEMISPHERE
If you live in Australia, New Zealand, or South Africa, look at the maps this side up. The main map shows the night sky as it appears when you look north at about 10:30 P.M. in the middle of the month. You should be able to see about as far north as the green line. The domed map above shows what you will see if you turn around and look south.

LOOKING SOUTH IN THE SOUTHERN HEMISPHERE
Sagittarius and Scorpius blaze high in the sky this month. The curved arc of stars forming the scorpion's deadly tail appears right in front of you. In the Milky Way, Crux has its long axis nearly horizontal and its short axis vertical.

DID YOU KNOW?
Antares, Scorpius's brightest star, is one of the biggest stars we can see in the sky. It is a red supergiant that may be as much as 500 times larger than the Sun.

JULY
FROM ARA TO SERPENS

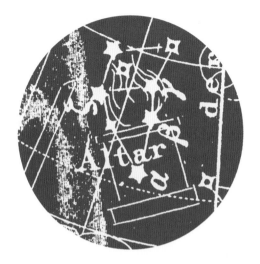

ARA, THE ALTAR
Ara is a small, far-southern constellation that lies just south of the curved "tail" of Scorpius. It represents the altar upon which the gods made their vows before going into battle, or maybe the altar on which the centaur (Centaurus) was going to sacrifice the wolf (Lupus). Ara lies on the edge of the Milky Way against a glorious background of stars.

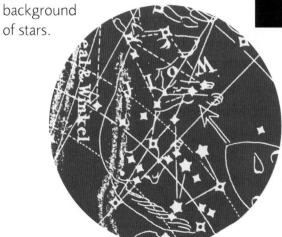

LUPUS, THE WOLF
Lupus is sandwiched between the two much more brilliant constellations Centaurus and Scorpius, so it is not difficult to find. It represents a wolf being carried on a pole by a centaur (Centaurus), maybe as a sacrifice on an altar (Ara). It lies on the opposite side of the Milky Way from Ara.

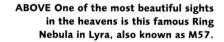

ABOVE One of the most beautiful sights in the heavens is this famous Ring Nebula in Lyra, also known as M57.

LYRA, THE LYRE

Tiny Lyra is dominated by the brilliant star called Vega. The constellation represents the musical instrument invented by Hermes (Mercury), the winged messenger of the gods. The constellation is also often drawn as a bird, because that is how Arabian astronomers saw it. The name Vega means something like "swooping eagle" in Arabic.

Vega is the fifth brightest star in the whole heavens. It is one of the three bright stars that make up the "summer triangle" (see page 60).

OPHIUCHUS, THE SERPENT BEARER

Ophiuchus is a large constellation representing a figure with a snake (the constellation Serpens) coiled round his body. He is identified with Asclepius, the Greek god of medicine, who was a son of Apollo and was taught the art of healing by the learned centaur Chiron.

The constellation contains a surprising number of globular clusters, those great closely packed balls of stars. You may spot two of them (M10 and M12) through binoculars in the relatively empty center of the constellation. Binoculars will also show stunning starscapes in the southern part of the constellation, where it meets the Milky Way.

The Milky Way lies in midskies this month and next, and is well placed for observation. Find out more about it on the next page.

SERPENS, THE SERPENT

Serpens is an oddity among the constellations because it is divided in two, by Ophiuchus. The head part of the serpent, or snake, is called Serpens Caput, from the Latin word

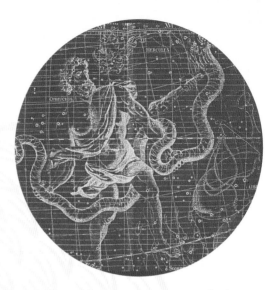

for "head." The tail part is called Serpens Cauda, from the Latin word for "tail."

The serpent represents a snake that the serpent-bearer (the Greek god of medicine Asclepius) once killed. But it was restored to life when another snake placed a herb on it. Asclepius then used the same herb to bring the dead back to life. And the snake became the symbol for healing.

Both the head and the tail have interesting objects. Not far from the brightest star Alpha (α) in the head is M5, a fine globular cluster well seen in binoculars. And in the tail on the edge of the Milky Way is M16, an open star cluster that is set in the Eagle Nebula.

THE MILKY WAY—
THE ARCH OF COUNTLESS STARS

On most clear nights when it's really dark, you can see a faint band of light across the sky. We call it the Milky Way. The band passes through certain constellations, so you'll see different parts of it, in different positions in the sky, at different times of the year. Among the main constellations through which the Milky Way passes are Perseus, Cassiopeia, Cygnus, Aquila, Sagittarius, Scorpius, Centaurus, Crux, Vela, Puppis, and Gemini. With the band, it makes these constellations look truly spectacular when viewed through binoculars.

STARS WITHOUT NUMBER

What exactly is the Milky Way? If you look at it through binoculars, you'll find out. You'll see that it is made up of a mass of stars, which seem to be packed tightly together.

Actually, these stars are sprinkled about in space like the other stars in the sky. They appear packed together because we live inside a disk of stars.

When we look along the disk, we are looking over great distances in space, and therefore see masses of stars. When we look out of the sides of the disk, we are looking over quite a short distance and therefore see fewer stars.

THE MILKY WAY GALAXY

What is this disk of stars we live in? It is a great star island in space—a galaxy (see page 80). We call it the Milky Way Galaxy, or just the Galaxy.

The Galaxy is not exactly shaped like a flat disk—it has a big bulge at the center. And the stars in the flatter, outer part of the disk are not spread out evenly. They gather together in streams, or "arms." These arms spiral out from the bulge at the center. The Galaxy is an example of a common kind of galaxy, called a spiral.

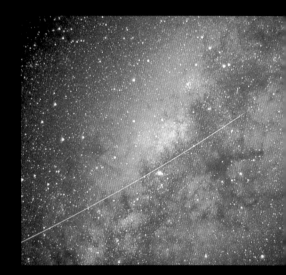

ABOVE A satellite streaks across the southern Milky Way.

Like everything in the Universe, the Galaxy is moving. It travels through space like all the other galaxies. As it does so, it spins round slowly, taking hundreds of millions of years to spin around once. Viewed from a great distance, the Galaxy would look like a flaming pinwheel firework.

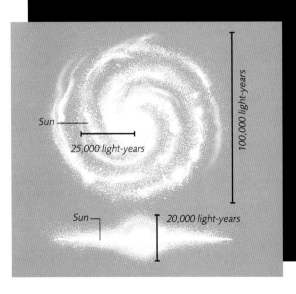

LEFT Dimensions of our home galaxy, the Milky Way.

THE MILKY WAY

LEFT One of the most beautiful starscapes in the heavens, this is the Sagittarius star cloud in the far southern Milky Way.

BELOW A side view of our galaxy, composed from satellite images.

THE GALAXY'S VITAL STATISTICS

The Galaxy is huge—it is bigger than we can ever imagine. It contains as many as 200 billion stars—many like the Sun, many more that are bigger and brighter, and many others that are smaller and dimmer. From one side to the other, the Galaxy measures 100,000 light-years. The Sun lies about 25,000 light-years from the center.

ARMS AND THE BULGE

Even though we live inside the Galaxy, astronomers have been able to find out how it's made up. They have done this using radio telescopes—telescopes that "tune" into the radio waves that the sky gives out. They find that the Galaxy has two main arms and parts of others.

The main arms are called the Sagittarius and Perseus Arms, after the constellations found on them. The Sun and familiar constellations such as Taurus, Orion, and Crux lie on the Orion, or Local, Arm.

The spiral arms contain a great deal of gas and dust, in which stars are continually being born. The bulge, on the other hand, does not, so few stars are born there. This means that, in general, the arms contain younger stars; and the bulge older ones.

The spiral arms contain many clusters of young stars, called open clusters.

The bulge has clusters of much older stars circling round it. They are called globular clusters. To find out more about clusters, see page 90.

AUGUST SKIES

The three bright stars of the "summer triangle" are still prominent—high in the Northern Hemisphere, lower (and upside-down, of course) in the Southern. Altair appears in midsky on the north–south line (or "meridian") in both hemispheres. The "summer triangle" straddles the Milky Way, which is worth inspecting here. It is highly fragmented, with a very clear split caused by the near-starless Cygnus Rift. This region is not really a starless part of the Milky Way, but an area where large amounts of thick dust obscure the light of the distant stars.

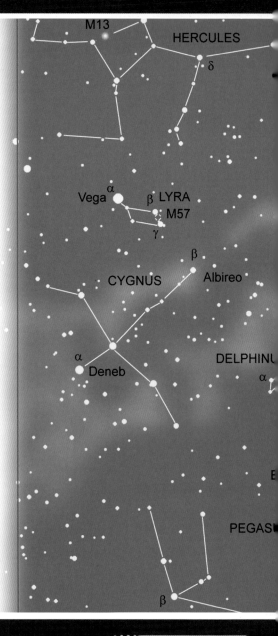

IN THE NORTHERN HEMISPHERE

If you live in the United States, Canada, or Europe, look at the maps this side up. The main map shows the night sky as it appears when you look south at about 10:30 P.M. in the middle of the month. You should be able to see about as far south as the red line. The domed map above shows what you will see if you turn around and look north.

LOOKING NORTH IN THE NORTHERN HEMISPHERE

Old friends reappear this month—bright Arcturus low down in the west, and Capella just above the horizon nearly in front of you. Close by, Perseus has begun to climb. Cassiopeia is now level with Polaris in midsky, and the Big Dipper is quickly descending.

EYE SPY

Look high overhead to see the three beacon stars of the "summer triangle."

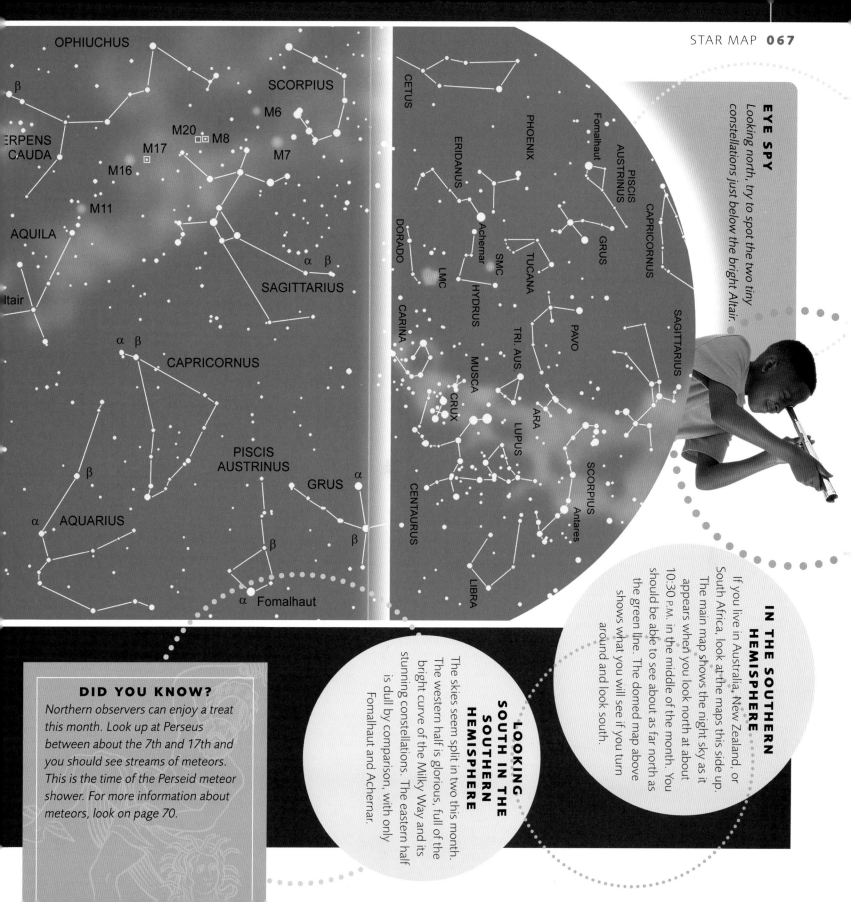

STAR MAP 067

EYE SPY
Looking north, try to spot the two tiny constellations just below the bright Altair.

IN THE SOUTHERN HEMISPHERE
If you live in Australia, New Zealand, or South Africa, look at the maps this side up. The main map shows the night sky as it appears when you look north at about 10:30 P.M. in the middle of the month. You should be able to see about as far north as the green line. The domed map above shows what you will see if you turn around and look south.

LOOKING SOUTH IN THE SOUTHERN HEMISPHERE
The skies seem split in two this month. The western half is glorious, full of the bright curve of the Milky Way and its stunning constellations. The eastern half is dull by comparison, with only Fomalhaut and Achernar.

DID YOU KNOW?
Northern observers can enjoy a treat this month. Look up at Perseus between about the 7th and 17th and you should see streams of meteors. This is the time of the Perseid meteor shower. For more information about meteors, look on page 70.

AUGUST
FROM AQUILA TO SAGITTARIUS

AQUILA, THE EAGLE

Aquila is one of the pair of "birds" that "fly" along the Milky Way this month (the other is Cygnus). It represents the eagle that carried the thunderbolts hurled by Zeus.

Aquila is a fine constellation, which contains the brilliant star Altair. This star is one of the trio that form the "summer triangle."

Through binoculars, Altair and its close companions Beta (β) and Gamma (γ) make a nice group. Compare brilliant white Altair with distinctly orange Gamma. Look also at the little arc of stars around Lambda (λ). Close by you'll find a star cluster (M11) that is called the Wild Duck.

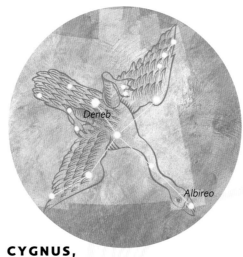

CYGNUS, THE SWAN

Cygnus is another constellation that lives up to its name. You need little imagination to see in this pattern of stars a swan in flight, with neck extended and wings outstretched. Cygnus is also called the Northern Cross. It represents one of the disguised forms that Zeus used when he pursued beautiful women. Among them was Leda, Queen of Sparta, who laid eggs from which were born the heavenly twins Castor and Pollux (in Gemini).

The brightest star in Cygnus is Deneb, a word meaning "tail" in Arabic. It is one star in the "summer triangle." The next brightest star, Albireo, marks the swan's beak. Try to look at Albireo in a telescope. You'll find that it is a binary star—one star is blue, the other yellow. They look lovely together.

Being in the Milky Way, Cygnus is rich in star clusters and nebulas. A glowing mass of gas near Deneb bears an uncanny resemblance to the continent of North America and is called the North America Nebula (NGC 7000).

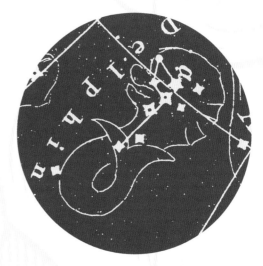

DELPHINUS, THE DOLPHIN

Delphinus is a tiny constellation, but it is a real gem to look at through binoculars. You can easily imagine its compact pattern of stars as a leaping dolphin. It represents the dolphin that acted as the messenger of the sea god Poseidon.

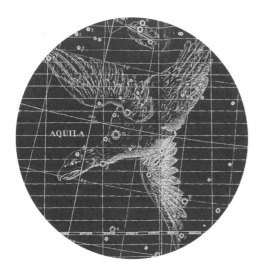

RIGHT The beautiful North America Nebula in Cygnus, which looks uncannily like the continent of North America.

SAGITTA, THE ARROW

Sagitta is another tiny, compact constellation, resembling an arrow in flight. Look for it north of bright Altair in Aquila. Its four main stars look lovely through binoculars. The Milky Way looks glorious here, too.

SAGITTARIUS, THE ARCHER

Sagittarius is a brilliant constellation that lies in the Milky Way south of Aquila. It is a constellation of the zodiac (pages 54–55). It represents the son of Pan, Crotus, who was said to have invented archery. He is usually depicted as a centaur, aiming his arrow at the heart of the scorpion (Scorpius).

Sagittarius is truly spectacular, containing an abundance of nebulas and star clusters. This is not surprising, however, because it lies in the densest part of the Milky Way, in the direction of the center of our Galaxy. These objects include the spectacular Lagoon, Omega, and Trifid Nebulas. They can be spotted in binoculars, but long-exposure photographs are needed to reveal their full beauty.

METEORS— STARS THAT FALL FROM THE SKY

During the second week in August every year, you can see many bright streaks in the constellation Perseus every night. It looks as if stars are falling out of the sky by the hundreds. But these "falling stars" are not stars at all. The bright streaks you see are trails made when specks of rock burn up in the air, high above the ground. We call them meteors. The brightest meteors happen when larger rocky lumps burn. Also, sometimes pieces don't burn up completely and fall all the way to the ground. We call these pieces meteorites.

BOMBARDMENT FROM SPACE

The space between the planets and around Earth is full of bits of rock and metal too. When these bits get close enough, Earth's gravity attracts them, and they begin to fall toward our planet.

These bits travel very fast—at speeds up to 150,000 miles (250,000 km) per hour. When they hit Earth's atmosphere, friction (rubbing) with the air particles heats them up. They quickly become so hot that they catch fire and burn up, leaving a fiery trail behind, which we see as a meteor.

SHOWERS OF METEORS

You will probably see a few meteors every night on which you go stargazing, once your eyes have become used to the dark. Around five an hour is about average. You'll find that they will come from any part of the sky.

At certain times of the year, however, you may see dozens, even hundreds, of meteors in an hour. We call this a meteor shower. In a shower, nearly all the meteors come from the same part of the sky.

PERSEIDS AND LEONIDS

In August, as we have mentioned, a shower of meteors comes from the direction of Perseus. We call it the Perseid meteor shower.

You can see one of the best of the other regular showers in mid-November, coming from Leo. It is called the Leonid shower.

Why do showers of meteors take place at certain times of the year? The answer is because of comets (page 22). When comets travel in toward the Sun, they leave trails of dust behind in space. When Earth comes along, this dust rains down into the atmosphere and causes meteor showers.

LEFT A meteor streaks across a photograph taken of a distant galaxy.

METEORS

CREATING CRATERS

Large meteorites hit the ground with tremendous force and dig out big holes, or craters. About 50,000 years ago, a meteorite weighing hundreds of thousands of tons smashed into the ground in what is now the Arizona Desert. It blasted out the famous Meteor Crater, near the town of Winslow. About 575 feet (175 meters) deep, Meteor Crater measures more than 3,900 feet (1,200 meters) across.

ABOVE
Most craters on the Moon were created by the impact of meteorites.

LOOKING AT METEORITES

Many meteorites have been found all over the world. Most are small, but a few are big. One found in Namibia in southwest Africa weighs about 60 tons.

Most meteorites found are made up of rocky material and are called stony meteorites, or stones. Most of the others are made up of metal and are called iron meteorites, or irons. A few are a mixture of stone and metal.

ABOVE During a meteor shower, hundreds of meteors may be seen every hour.

RIGHT Astronomers think that this meteorite may have come from Mars.

SEPTEMBER SKIES

Days are noticeably shortening in the Northern Hemisphere and lengthening in the Southern. On or around September 23, day and night are of equal length. It is the time of the autumnal, or fall, equinox. On that date, the Sun is located directly above the Equator and is traveling from the Northern to the Southern Hemisphere. This signals the beginning of fall in the Northern Hemisphere and of spring in the Southern. The "summer triangle" stars are slipping away west, and Pegasus is taking their place. (Compare with March Skies, page 40.)

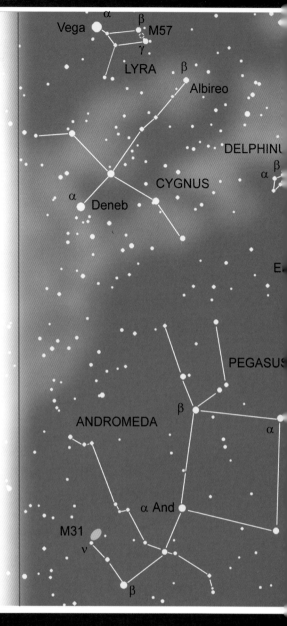

IN THE NORTHERN HEMISPHERE

If you live in the United States, Canada, or Europe, look at the maps this side up. The main map shows the night sky as it appears when you look south at about 10:30 P.M. in the middle of the month. You should be able to see about as far south as the red line. The domed map above shows what you will see if you turn around and look north.

LOOKING NORTH IN THE NORTHERN HEMISPHERE

This month the Big Dipper is near its lowest point in the northern heavens, being nearly parallel with the horizon. Its two pointer stars point almost straight up. In the east, Capella has now been joined by Aldebaran, the red eye of the bull.

EYE SPY

Looking north, is Arcturus still above the western horizon where you live?

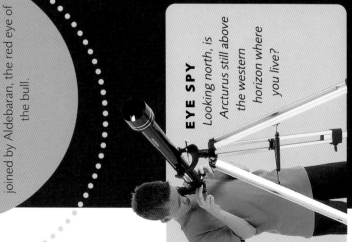

STAR MAP 073

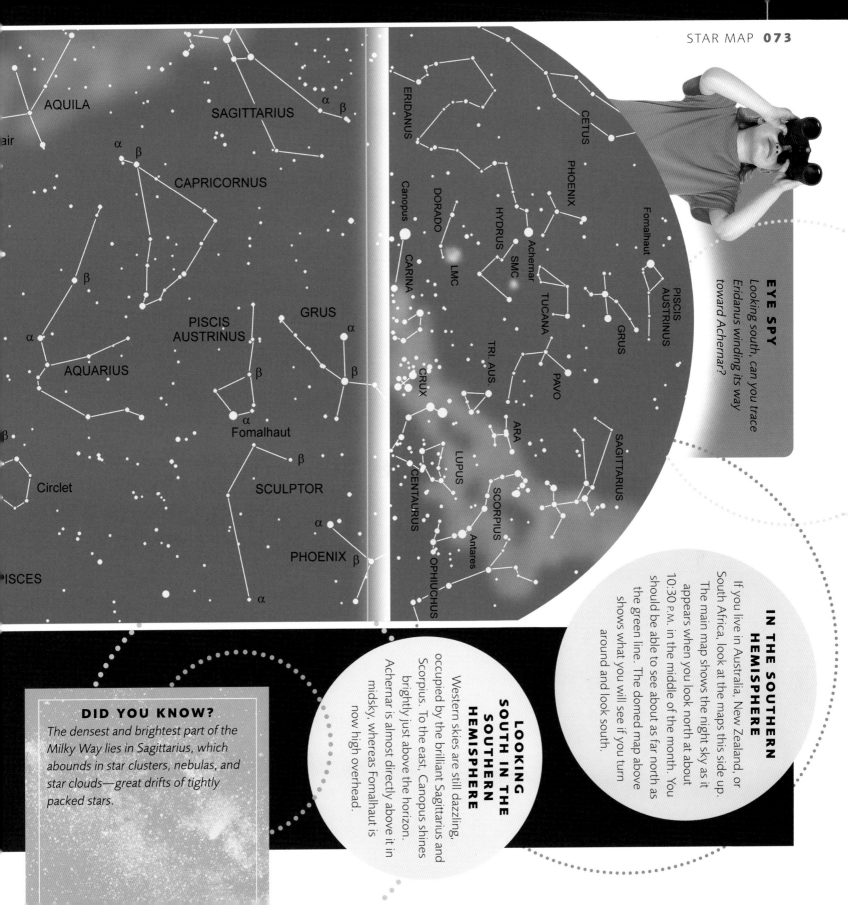

EYE SPY
Looking south, can you trace Eridanus winding its way toward Achernar?

IN THE SOUTHERN HEMISPHERE
If you live in Australia, New Zealand, or South Africa, look at the maps this side up. The main map shows the night sky as it appears when you look north at about 10:30 P.M. in the middle of the month. You should be able to see about as far north as the green line. The domed map above shows what you will see if you turn around and look south.

LOOKING SOUTH IN THE SOUTHERN HEMISPHERE
Western skies are still dazzling, occupied by the brilliant Sagittarius and Scorpius. To the east, Canopus shines brightly just above the horizon. Achernar is almost directly above it in midsky, whereas Fomalhaut is now high overhead.

DID YOU KNOW?
The densest and brightest part of the Milky Way lies in Sagittarius, which abounds in star clusters, nebulas, and star clouds—great drifts of tightly packed stars.

SEPTEMBER
FROM AQUARIUS TO PISCIS AUSTRINUS

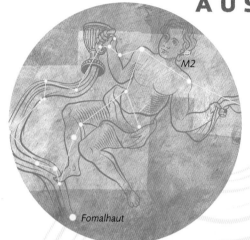

AQUARIUS, THE WATER BEARER

Aquarius is a sprawling constellation found in a "watery" region of the heavens, surrounded by the fishes (Pisces), whale (Cetus), sea-goat (Capricornus), and southern fish (Piscis Austrinus). It is a constellation of the zodiac (see pages 54–55).

The pattern of stars in Aquarius represents a youth pouring water from a jar into the mouth of the southern fish, marked by the bright star Fomalhaut in the constellation Piscis Austrinus. The youth is the handsome Ganymede, who was the son of King Tros (founder of Troy) and who became the cupbearer of the gods.

Aquarius is a relatively faint constellation, and can perhaps best be found by looking south from Pegasus, easily recognizable by its famous square. With binoculars, look for the fine globular cluster M2, which forms one corner of a right-angled triangle with Alpha (α) and Beta (β).

CAPRICORNUS, THE SEA GOAT

Capricornus is another faint constellation of the zodiac (see pages 54–55), close to Aquarius. Its shape is like that of a crooked triangle. It represents a strange creature with the tail of a fish and the body of a goat. The Greeks associated it with their god Pan, who was part man, and part goat.

You should be able to see with the naked eye that Alpha (α) is a double star. The two component stars, both yellow, are not really close together, but just happen to lie in the same direction in space.

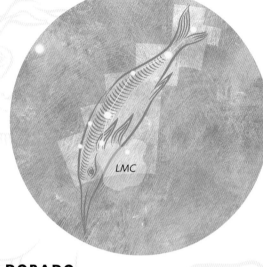

DORADO, THE SWORDFISH

Also sometimes known as the Goldfish, Dorado is a small far-southern constellation. It is most interesting because it contains one of the most striking features in southern skies, a large misty patch that looks like a wispy cloud. It is called the Large Magellanic Cloud (LMC). However, it is not a cloud—it is a neighboring galaxy, one of only three that we can see with the naked eye. Through binoculars you can see in the LMC a bright mass of gas, called the Tarantula Nebula because of its spidery shape (see picture opposite).

CONSTELLATIONS 075

GRUS, THE CRANE

Grus represents a bird in flight, with a long neck and outstretched wings. It is a "southern bird," like Pavo (see page 59). Can you spot the color difference between Alpha (α) and Beta (β)? Also, see if you can see separately the two stars that make up the double star Delta (δ). They look close together, but they are really far apart. Mu (μ) is a double star, too.

PISCIS AUSTRINUS, THE SOUTHERN FISH

The dominant star in this small southern constellation is Fomalhaut. It marks the mouth of the fish into which water is pouring from the water jar in Aquarius. Although Fomalhaut ranks as only the eighteenth brightest star in the heavens, it shows up well because it lies in a relatively dull part of the night sky.

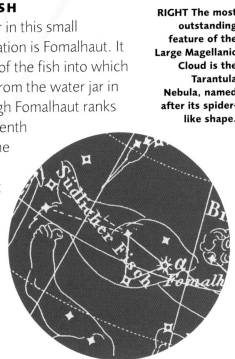

RIGHT The most outstanding feature of the Large Magellanic Cloud is the Tarantula Nebula, named after its spider-like shape.

THE CLOUDS OF MAGELLAN

The Large Magellanic Cloud and Small Magellanic Cloud are named for a Portuguese sailor, Ferdinand Magellan. In the early 1500s, he became one of the first Europeans to sail the southern seas, and would probably have used the Clouds to help him navigate.

OCTOBER SKIES

Pegasus occupies center stage this month, signaling that fall has arrived in the Northern Hemisphere and spring in the Southern. Its famous square of stars lies on the north–south line in both hemispheres. In the Northern Hemisphere, it is a very good time to observe the Andromeda Galaxy (M31) in the skies north of the Square. It seems astonishing that we can see this object with the naked eye 15 million million million miles (25 million million million kilometers) away. Southern observers have a poor view of the galaxy, because it remains low down near the horizon.

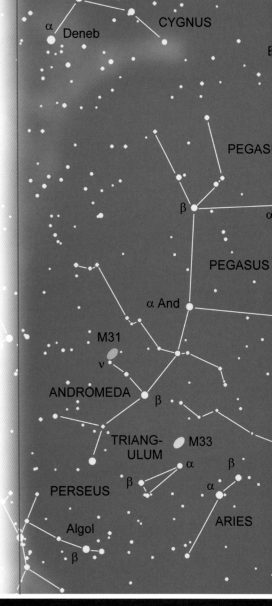

IN THE NORTHERN HEMISPHERE

If you live in the United States, Canada, or Europe, look at the maps this side up. The main map shows the night sky as it appears when you look south at about 10:30 P.M. in the middle of the month. You should be able to see about as far south as the red line. The domed map above shows what you will see if you turn around and look north.

LOOKING NORTH IN THE NORTHERN HEMISPHERE

In the west, two stars of the "summer triangle," Deneb and Vega, are descending, reminding us that summer is long gone. In the east, Gemini's Castor and Pollux are rising, one above the other. They line up with Capella in midsky.

EYE SPY

Looking south, look for Fomalhaut low down near the horizon right in front of you.

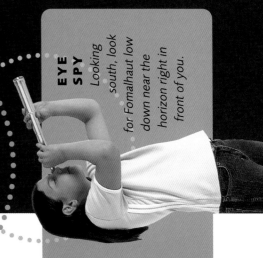

STAR MAP 077

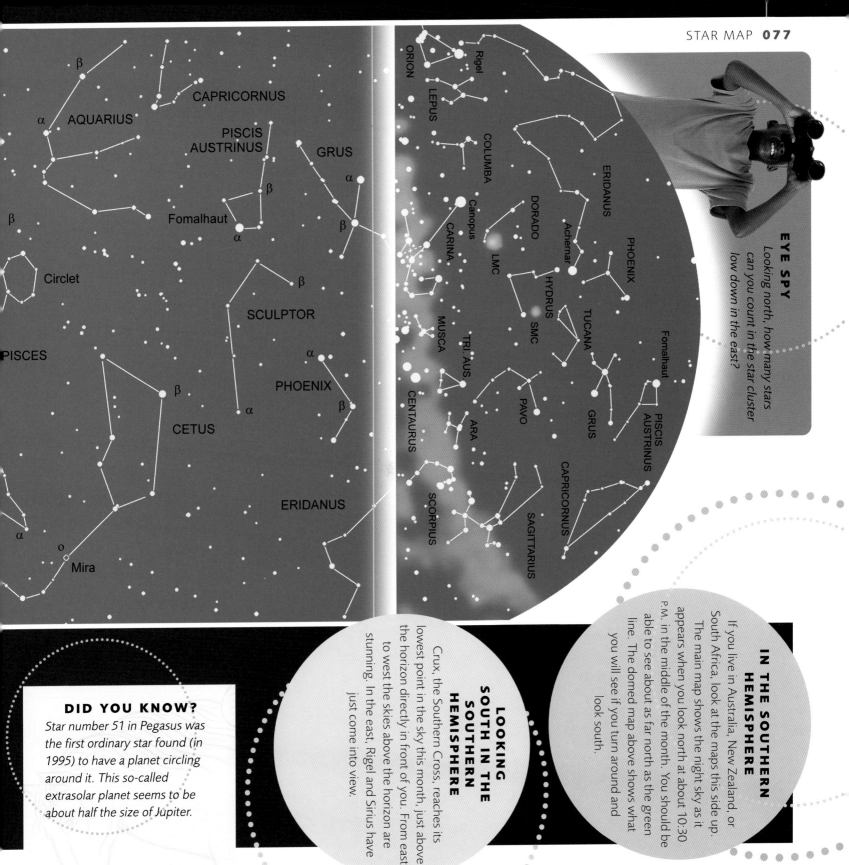

EYE SPY
Looking north, how many stars can you count in the star cluster low down in the east?

IN THE SOUTHERN HEMISPHERE
If you live in Australia, New Zealand, or South Africa, look at the maps this side up. The main map shows the night sky as it appears when you look north at about 10:30 P.M. in the middle of the month. You should be able to see about as far north as the green line. The domed map above shows what you will see if you turn around and look south.

LOOKING SOUTH IN THE SOUTHERN HEMISPHERE
Crux, the Southern Cross, reaches its lowest point in the sky this month, just above the horizon directly in front of you. From east to west the skies above the horizon are stunning. In the east, Rigel and Sirius have just come into view.

DID YOU KNOW?
Star number 51 in Pegasus was the first ordinary star found (in 1995) to have a planet circling around it. This so-called extrasolar planet seems to be about half the size of Jupiter.

OCTOBER
FROM ANDROMEDA TO TUCANA

ANDROMEDA

Andromeda represents one of the most tragic figures in Greek mythology, whose story links many of the surrounding constellations. Andromeda was a fair maiden offered as a sacrifice by her father (King Cepheus) as a result of boastful claims by his wife (Queen Cassiopeia).

The queen boasted that she was more beautiful than any other woman, which upset the beautiful sea nymphs. So the sea god Poseidon (Neptune) sent a monster (Cetus) to lay waste the king's lands. The king was told it would go away if he would sacrifice his daughter to it.

So poor Andromeda was chained to rocks on the seashore, and the monster rushed in to devour her. But the hero Perseus came along in the nick of time, and killed the monster, saved Andromeda, and claimed her for his bride.

The astronomical highlight in Andromeda is a misty patch near the star Nu (ν). It is identified as M31 and was once thought to be a nebula among the stars in the sky. But we now know it is a separate star system, or galaxy, far beyond the stars. For more information about the Andromeda Galaxy and other galaxies, look over the page.

RIGHT One of the globular clusters that surround the center of the Andromeda Nebula. It contains many thousands of stars.

CONSTELLATIONS 079

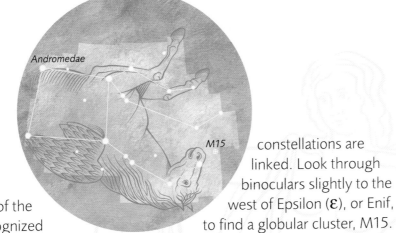

PEGASUS, THE FLYING HORSE

Pegasus is one of the most easily recognized constellations, not because it looks like a horse with wings, but because it features an almost perfect square of bright stars, the Square of Pegasus.

Pegasus represents the flying horse that sprang from the blood that gushed from the body of Medusa, the terrifying Gorgon, after she had been killed by the hero Perseus.

The three brightest stars in Pegasus mark three corners of the distinctive square. The other corner is marked by Andromeda's brightest star (Alpha Andromedae) because the two constellations are linked. Look through binoculars slightly to the west of Epsilon (ε), or Enif, to find a globular cluster, M15.

PISCES, THE FISH

Pisces is a scattered constellation of the zodiac (pages 54–55), difficult to make out because its stars are so faint. It is best found by looking for a circle of stars (the "Circlet"), just south of the Square of Pegasus. This circle marks the southernmost of the two fishes that make up the constellation. The fishes represent Aphrodite (Venus), the goddess of love, and her son Eros (Cupid). It recalls the time when they had to jump into a river to escape the clutches of the monster Typhon.

TUCANA, THE TOUCAN

Tucana is one of the "southern birds" (see page 59), representing the colorful big-beaked bird of tropical America. This small constellation has two quite outstanding features. One is 47 Tucanae, which looks like a fuzzy star but is actually a huge globular cluster. You can easily see it with the naked eye; but, for a treat, look at it through binoculars. Tucana's other highlight is a misty patch called the Small Magellanic Cloud (SMC), which is a neighboring galaxy (see also pages 74–75).

SPRING HAS SPRUNG

The Sun crosses over the celestial equator in Pisces, moving from the Southern to the Northern Hemisphere, on about March 20 each year. This signals the beginning of spring in the Northern Hemisphere and the beginning of fall in the Southern.

LEFT Just the head and forelimbs of Pegasus appear in the sky.

GALAXIES—STAR ISLANDS IN SPACE

On a clear, dark night, find Cassiopeia and the Square of Pegasus. Then look about halfway between the two. You'll see a faint misty patch, close to Andromeda's star Nu (ν). Astronomers once thought it was a nebula, or cloud of gas among the stars, but we now know it is a completely separate star system, far beyond the stars in our skies. It is one of the billions of star islands in space that we call galaxies. The Andromeda Galaxy is one of our closest galaxy neighbors in space—but it still lies 2.5 million light-years away.

SOUTHERN CLOUDS

The Andromeda Galaxy, also known as M31, can best be seen by observers in the Northern Hemisphere. Observers in the Southern Hemisphere, on the other hand, can see other galaxies with the naked eye. These include the Large and Small Magellanic Clouds, which are much smaller star systems than the Andromeda Galaxy, but they look bigger and brighter because they lie much closer to us. Nearest is the Large Magellanic Cloud (LMC), at a distance of only about 160,000 light-years.

The Sun and all the other stars in our skies belong to our home galaxy, which we call the Milky Way Galaxy, or the Galaxy. To find out more about it, turn to page 64.

SHAPING UP THE GALAXIES

Our Galaxy is shaped like a giant pinwheel and so is the Andromeda Galaxy. They both have a thick bulge of stars in the center, with other stars strung out on curved "arms" that spiral out from the center.

There are many galaxies like this in the Universe, and we call them spirals. The Andromeda Galaxy is one of the biggest spirals we know. It contains twice as many stars (400 billion) as our own Galaxy but is half as wide again (150,000 light-years).

Many other galaxies do not have spiral arms, but have a round or elliptical (oval) shape. We call them ellipticals. Many ellipticals are much smaller than our Galaxy, but a few are much larger. The biggest measures up to 500,000 light-years across—five times the size of our Galaxy.

LEFT When you peer deep into space, you see galaxies galore. These galaxies, spied by the Hubble Space Telescope, lie billions of light-years away.

GALAXIES

LEFT A famous galaxy neighbor, the Great Spiral in Andromeda, M31.

HYPERACTIVE!
A few galaxies pour out much more energy into space than usual. Sometimes they give off this extra energy as light; other times they give it off as invisible rays, such as radio waves and X-rays. We call these highly energetic objects active galaxies. They include radio galaxies, quasars, and blazars. Astronomers believe that these galaxies have massive black holes at their centers, which produce their exceptional energy.

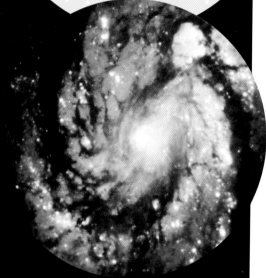

ABOVE The galaxy M100 is one of thousands that make up the Coma cluster of galaxies.

Some galaxies have no regular shape at all. We call them irregulars. The Magellanic Clouds are irregulars. In general, irregulars are small—some contain only a few million stars.

CLUSTERING TOGETHER
Our Galaxy, the Andromeda Galaxy, and the Magellanic Clouds all belong to a small cluster of about 30 galaxies in our part of the Universe called the Local Group. There is just one other spiral in the Group—the Triangulum Galaxy, or M33. Most other galaxies are small ellipticals and irregulars.

Elsewhere in the Universe, galaxies gather together by their hundreds and even thousands to form clusters. More than 2,000 galaxies are found in a cluster in the constellation Virgo.

In turn, clusters of galaxies themselves gather together to form incredibly big superclusters. And all the great superclusters link up to form our great Universe.

NOVEMBER SKIES

As Pegasus sinks in the west, the spectacular constellations of winter in the Northern Hemisphere and summer in the Southern are moving in from the east. Taurus is well placed for observation in both hemispheres during this and the coming months. The bull's orange "eye" Aldebaran and its two famous clusters, the Hyades and Pleiades, are always a delight to watch. But much of the sky is relatively dull, occupied by faint, "watery" constellations, such as the meandering river Eridanus, the sea monster Cetus, and the fishes Pisces.

IN THE NORTHERN HEMISPHERE

If you live in the United States, Canada, or Europe, look at the maps this side up. The main map shows the night sky as it appears when you look south at about 10:30 P.M. in the middle of the month. You should be able to see about as far south as the red line. The domed map above shows what you will see if you turn around and look north.

LOOKING NORTH IN THE NORTHERN HEMISPHERE

The Milky Way arcs from east to west across the sky this month, its high point being marked by Cassiopeia, appearing as an "M" rather than a "W." In the west, Cygnus is well placed for viewing. It's the last time we'll see the trio Deneb, Vega, and Altair together for some time.

EYE SPY

Looking north, spot the triangle of bright stars in the west.

STAR MAP 083

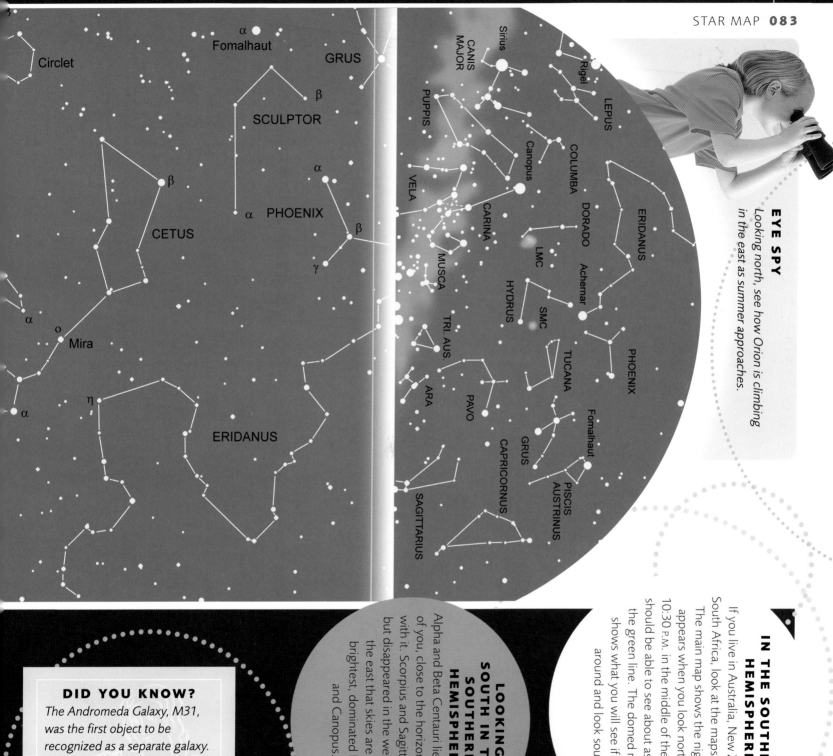

EYE SPY
Looking north, see how Orion is climbing in the east as summer approaches.

IN THE SOUTHERN HEMISPHERE
If you live in Australia, New Zealand, or South Africa, look at the maps this side up. The main map shows the night sky as it appears when you look north at about 10:30 P.M. in the middle of the month. You should be able to see about as far north as the green line. The domed map above shows what you will see if you turn around and look south.

LOOKING SOUTH IN THE SOUTHERN HEMISPHERE
Alpha and Beta Centauri lie right in front of you, close to the horizon and parallel with it. Scorpius and Sagittarius have all but disappeared in the west, and it is in the east that skies are now the brightest, dominated by Sirius and Canopus.

DID YOU KNOW?
The Andromeda Galaxy, M31, was the first object to be recognized as a separate galaxy. U.S. astronomer Edwin Hubble proved that it lay far beyond our own Galaxy in 1923.

NOVEMBER
FROM ARIES TO TRIANGULUM

BELOW This sparkling collection of stars can be found in Crux. It is an open cluster, aptly named the Jewel Box.

ARIES, THE RAM
Aries is a small constellation of the zodiac (see pages 54–55), with just two fairly bright stars. They can best be located by first finding the Pleiades star cluster and then looking due west toward the Square of Pegasus. Aries represents the ram with the Golden Fleece, which Jason and the Argonauts were sent to look for, and eventually found after many exciting adventures.

CETUS, THE WHALE
Cetus is a huge constellation, the fourth largest in the heavens. It represents the sea monster that ravaged King Cepheus's kingdom and was killed by Perseus when it was about to devour the fair Andromeda. For more about the story, see Andromeda (see page 78).

Astronomically, the most interesting star is Omicron (O), also known as Mira (meaning the "wonderful one"). Mira is distinctly reddish in color and varies widely in brightness.

CRUX, THE SOUTHERN CROSS

The unmistakable Southern Cross is the most famous of all the southern constellations. For most of Australia and New Zealand it is circumpolar and visible every night. This month southern observers can see it low down near the southern horizon. Most northern observers can never see it.

Ancient astronomers regarded Crux as part of the legs of the centaur (Centaurus). The two brightest stars in that constellation—Alpha (α) and Beta (β)—act as useful pointers to the Cross (see also pages 44 and 45).

Of the four main stars that make up the Cross, notice that Gamma (γ) is much redder than the other three. Near Beta is a colorful star cluster called the Jewel Box. It looks lovely through binoculars and sensational in a small telescope.

PHOENIX, THE PHOENIX

Phoenix is one of the "southern birds" (page 59). It is named after the fabulous bird that lived for hundreds or thousands of years, and then burned itself to death on a funeral pyre. A reborn phoenix emerged from the ashes. The constellation is probably best found by looking north of Eridanus's bright star Achernar.

TRIANGULUM, THE TRIANGLE

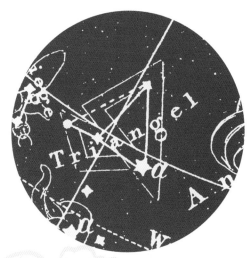

Triangulum is a neat little northern constellation, probably best located by looking east of the Square of Pegasus. To some ancient astronomers, its shape suggested the delta of the River Nile, to others the island of Sicily. Its main claim to fame is a close galaxy, M33, also called Triangulum. This galaxy has widespread open arms and can be seen through binoculars in a really dark sky. Find it by looking westward from Alpha (α).

WONDERFUL VARIABLES

Astronomers find many stars in the heavens like Cetus's Mira that vary dramatically in brightness over fairly long periods (11 months for Mira). They are called long-period, or Mira, variables. They appear to be enormous red giant stars that pulsate, first swelling up, and then shrinking in size. As they do so, they first brighten, and then fade.

DECEMBER SKIES

The spectacular constellations of winter in the Northern Hemisphere and summer in the Southern fill eastern skies this month. Canis Major, Orion, and Gemini are the dominant ones, with their brilliant stars shining like heavenly beacons. They will stay with us all winter (summer) long.

On or around December 20 is the winter "solstice." In the Northern Hemisphere, this is when the Sun reaches its lowest point in the skies, and days are shortest. In the Southern Hemisphere, the Sun reaches its highest point in the skies on this date, and days are longest.

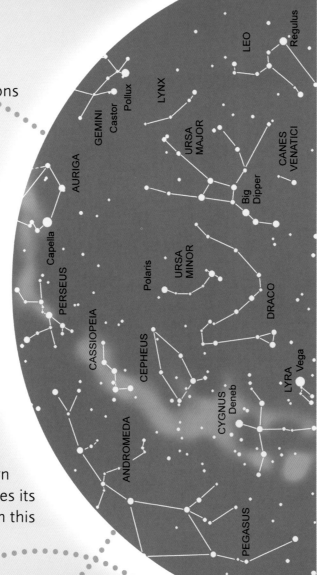

IN THE NORTHERN HEMISPHERE

If you live in the United States, Canada, or Europe, look at the maps this side up. The main map shows the night sky as it appears when you look south at about 10:30 P.M. in the middle of the month. You should be able to see about as far south as the red line. The domed map above shows what you will see if you turn around and look north.

LOOKING NORTH IN THE NORTHERN HEMISPHERE

The handle of the Little Dipper now hangs down almost vertically from Polaris. Toward the east, Cygnus is sinking below the horizon, but its tail star Deneb is still on show. Toward the west, Regulus in the lion's head has risen. Capella remains high overhead.

EYE SPY

Looking south, look for brilliant Sirius low down toward the east.

STAR MAP 087

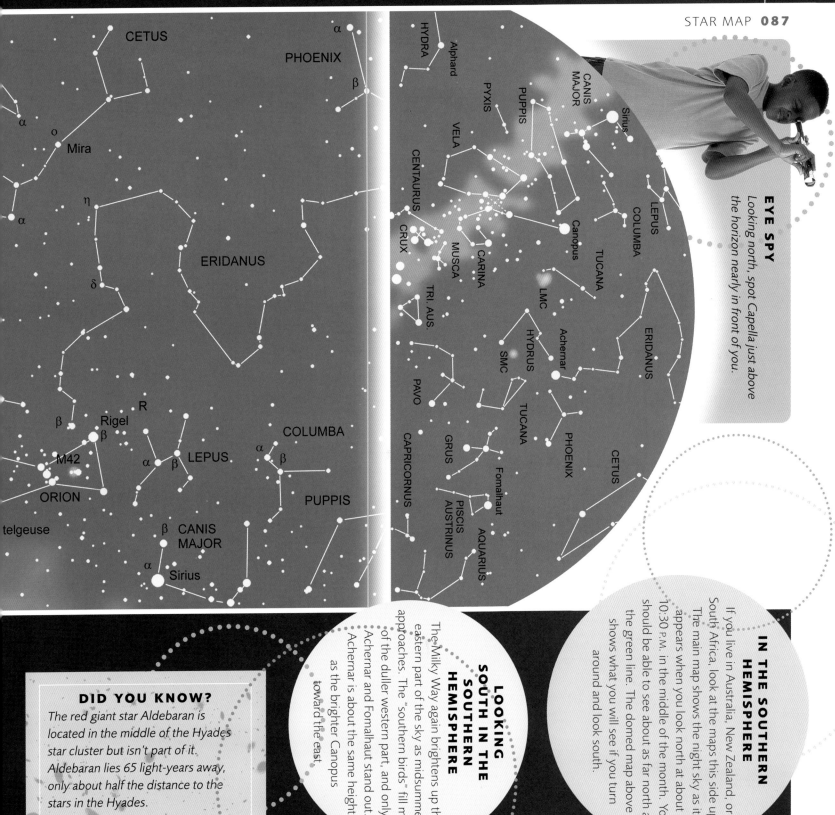

EYE SPY
Looking north, spot Capella just above the horizon nearly in front of you.

IN THE SOUTHERN HEMISPHERE
If you live in Australia, New Zealand, or South Africa, look at the maps this side up. The main map shows the night sky as it appears when you look north at about 10:30 P.M. in the middle of the month. You should be able to see about as far north as the green line. The domed map above shows what you will see if you turn around and look south.

LOOKING SOUTH IN THE SOUTHERN HEMISPHERE
The Milky Way again brightens up the eastern part of the sky as midsummer approaches. The "southern birds" fill much of the duller western part, and only Achernar and Fomalhaut stand out. Achernar is about the same height as the brighter Canopus toward the east.

DID YOU KNOW?
The red giant star Aldebaran is located in the middle of the Hyades star cluster but isn't part of it. Aldebaran lies 65 light-years away, only about half the distance to the stars in the Hyades.

DECEMBER
FROM COLUMBA TO TRIANGULUM AUSTRALE

COLUMBA, THE DOVE

Although Columba is a faint constellation, it can easily be picked up because it lies close to Canis Major. Added to the constellations in the late 1500s, it represents the dove that Noah released to search for dry land after the Flood. The star Mu (μ) is interesting because it travels very much faster than almost any other star, at a speed of approximately 250,000 miles (400,000 km) per hour. It is one of only a handful of so-called runaway stars.

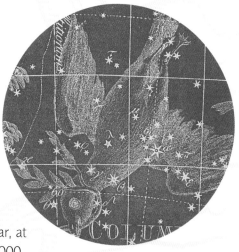

ERIDANUS

Eridanus is the longest and fourth largest constellation, which winds its way southward from Orion to near the south celestial pole. It represents a river—the Babylonians thought of it as the Euphrates; the Egyptians, as the Nile; and the Greeks, as the Po (in Italy). Eridanus "rises" near Orion's bright star Rigel. The constellation's second brightest star Beta (β) marks its source. It meanders leisurely south to its mouth, marked by its brightest star, Achernar (Arabic for "end of the river").

Astronomically, Eta (η) Eridani is one of the most interesting stars. It is one of the closest stars to us that is like the Sun, and it may well have a planetary system like ours. Could there be another Earth there?

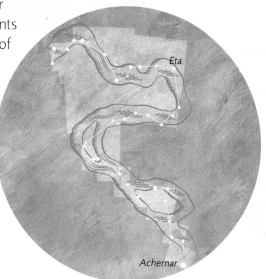

PERSEUS

Perseus is a splendid constellation that straddles the northern Milky Way. It represents one of the most dashing Greek heroes, who vanquished all kinds of monsters, such as the Gorgon Medusa. This hideous woman had snakes for hair and a stare that would turn people to stone. Perseus managed to get close enough to kill her because he had a helmet that made him invisible, and he looked at her only in the mirror-like surface of his shield. It was after he had killed the Gorgon that he saved Andromeda from the monster Cetus (see page 78).

Being in the Milky Way, Perseus boasts many fine starscapes and clusters. Its star Beta (β), or Algol, is particularly interesting. It varies noticeably in brightness every few days and is an example of a variable star called an eclipsing binary (see page 91).

CONSTELLATIONS

LEFT The famous Double Cluster in Perseus. This pair of small star clusters is set in the Milky Way and looks beautiful through binoculars or a small telescope.

TAURUS, THE BULL

Taurus is one of the richest constellations, full of interesting objects. It is a constellation of the zodiac (see pages 54–55). It represents the handsome bull into which Zeus changed himself when he pursued the beautiful Europa. She later gave birth to three children. One was Minos, who became King of Crete and started the cult of bull worship there.

Taurus's delights are many. Its noticeably reddish leading star Aldebaran marks the eye of the bull. Around it are a swarm of fainter stars, forming a cluster known as the Hyades. Farther north is an even more splendid cluster, the Pleiades. Find out more about the Pleiades and other star clusters over the page.

TRIANGULUM AUSTRALE, THE SOUTHERN TRIANGLE

Triangulum Australe is the southern equivalent of the north's Triangulum. Introduced in the late 1500s, it has brighter stars than Triangulum. Use Alpha and Beta Centauri to find it—it lies quite close to these stars on the edge of the Milky Way.

CLUSTERS—STARS THAT TRAVEL TOGETHER

Taurus is a spectacular constellation. Find Aldebaran, the star that marks the red eye of the bull, by tracing an imaginary line northwest through the three diagonal stars in the middle of Orion. Continue a little way past this star, and you'll see one of the most delightful sights in the heavens—the group of stars we call the Pleiades. It is one of the finest star clusters there is. How many stars can you count in the cluster? Some people say they can count seven, which is why it is also called the Seven Sisters.

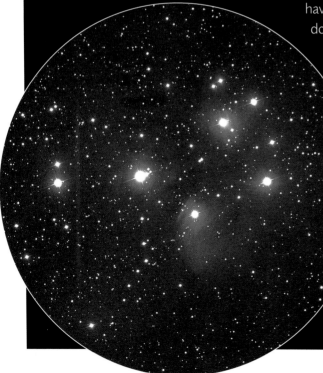

BELOW The best known of all open star clusters, the Pleiades, or Seven Sisters, in Taurus. In all, this cluster contains several hundred stars.

SEEING DOUBLE

Our local star, the Sun, travels through space by itself, but more than two-thirds of all stars travel through space with one or more companion stars. Almost half have just one companion. We call this a double, or binary, star, and each star a component. In such a system, the stars are linked by gravity and revolve around each other.

Sirius, the brightest star in the sky, is a double.

Some stars appear close together in the sky, and look like double stars, but in fact are not connected—they just happen to lie in the same direction in space. We call such stars optical doubles. An example is Mizar, the middle star in the handle of the Big Dipper (Plow). It forms an optical double with Alcor.

Some stars are made up of three or more component stars, and we call them multiple systems. A notable example is the third brightest star in the sky, Alpha Centauri. However, it is actually a system of three stars, one of which is called Proxima ("nearest") Centauri because it is the nearest star to us after the Sun. This star lies less than 4.3 light-years away.

STAR GROUPS

The stars that form the patterns of the constellations look as though they are traveling together through space. But this is in fact an illusion—they actually lie far apart. We see them together just because they happen to lie in the same direction in space.

However, there are large groups of stars that do travel together. The Pleiades is a brilliant example of what we call an open cluster. Altogether it

CLUSTERS 091

FAR LEFT The globular cluster M13 in Hercules contains hundreds of thousands of stars.

LEFT A small cluster of young, massive, and very hot stars (bottom right) can be found in the Tarantula Nebula. This nebula is a great stellar nursery in our galaxy neighbor, the Large Magellanic Cloud.

ECLIPSING BINARIES

About every two days or so, Perseus's star Algol dims for a few hours, then regains its usual brightness. It dims because it is a binary star. As the two stars revolve around each other, they periodically eclipse, or pass in front of each other as we see them. When the dimmer star eclipses the brighter one, Algol's overall brightness dips. When the dim star passes on, the overall brightness returns to normal. We know of many eclipsing binary stars like this.

contains several hundred stars, bound loosely together. They are all young, hot, and bright. All the open clusters we see in our Galaxy are made up of young, hot stars, and they are always found in the Galaxy's dusty "arms."

GLOBULARS

The most amazing star groups of all, however, consist of up to a million stars that are packed very close together in the shape of a ball or globe. We call them globular clusters. The stars in these clusters are mostly old, unlike the stars found in open clusters. Globular clusters are not found in the arms of the Galaxy, as open clusters are. They are found much farther away near the Galaxy's center.

WORDS TO REMEMBER

ACTIVE GALAXY
A galaxy that gives out extraordinary energy.

ASTEROIDS
Lumps of rock and metal circling the Sun, mainly between the orbits of Mars and Jupiter.

ASTROLOGY
An ancient belief that the heavenly bodies somehow affect human lives.

ASTRONOMY
The study of the heavens and the heavenly bodies.

ATMOSPHERE
The layer of gases around the Earth or other planets.

ATOM
The smallest bit of matter.

BILLION
One thousand million, or 1,000,000,000.

BINARY STAR
A double-star system in which the two stars circle around each other.

BLACK HOLE
A region of space that has such powerful gravity that not even light can escape.

BLAZAR
One kind of active galaxy.

CELESTIAL SPHERE
An imaginary dark sphere that appears to surround Earth.

CLUSTER
A group of stars or galaxies.

COMET
An icy lump that shines when it approaches the Sun.

CONSTELLATION
A collection of stars that seem to be grouped together in the sky.

CRATER
A pit in the surface of a planet or moon, usually dug out by a meteorite.

DOUBLE STAR
A star that looks like a single star but is actually two stars seen close together. Also called a binary star.

ECLIPSE
When one heavenly body passes in front of another and blots out its light. A lunar eclipse is an eclipse of the Moon (when the Earth passes in front of the Sun blocking the Sun's light to the Moon); a solar eclipse, an eclipse of the Sun (when the Moon passes in front of the Sun).

ELEMENTS
Known also as chemical elements. The basic "building blocks" of matter that cannot be broken down into simpler substances.

EVENING STAR
A first point of light to become visible in the night sky as it darkens.

EXTRASOLAR PLANET
A planet in orbit around another star.

FALLING STAR
A common name for a meteor.

GALAXY
One of billions of "star islands" in space; a giant collection of gas, dust, and millions, or billions, of stars.

GAS GIANT
In our Solar System, one of the planets Jupiter, Saturn, Uranus, and Neptune, which are huge and made up mainly of gas and liquid gas.

GRAVITY
The force that makes one body attract another.

HEAVENS
The night sky. The heavenly bodies are objects we see in the night sky.

HYDROGEN
The most common element in the Universe. Stars use it to make energy.

INTERPLANETARY
Between planets.

INTERSTELLAR
Between stars.

LIGHT-YEAR
A common unit for measuring distances in space. It is the distance that light travels in a year, about 6 million million miles (10 million million kilometers).

LUNAR
To do with the Moon.

MAGNITUDE
The brightness of a star.

WORDS TO REMEMBER

MARE (PLURAL MARIA)
A dark, dusty plain on the Moon, from the Latin word for "sea."

METEOR
A streak of light in the sky caused by a piece of rock burning up in the air.

METEORITE
A lump of rock or metal that has fallen to the Earth from outer space.

MILKY WAY
A pale band seen in the night sky, which is made up of millions of stars. The Milky Way Galaxy is our home galaxy, to which all the stars in the sky belong.

MORNING STAR
The last point of light to disappear as sunrise approaches.

MOON
A body that circles around a planet. The Moon is the Earth's only moon.

NEBULA
A cloud of gas and dust in space.

NUCLEAR REACTION
A process that involves the nuclei (centers) of atoms.

OBSERVATORY
A place where astronomers work and make their observations.

ORBIT
The path in space that one body follows when it circles around another.

PHASES
The apparent changes in the shape of the Moon during the month.

PLANET
One of nine bodies that circle around the Sun, including Earth. Extrasolar planets are planetary bodies that circle around other stars.

QUASAR
One kind of active galaxy, which looks like a star but is actually a bright and distant galaxy.

RADIO ASTRONOMY
A branch of astronomy in which astronomers study the radio waves coming from space.

RED GIANT
A large red star near the end of its life.

REFLECTOR
A telescope that uses mirrors to gather and focus starlight.

REFRACTOR
A telescope that uses lenses to gather and focus starlight.

SATELLITE
A small body that circles around a larger one; a moon. It is also the usual term for an artificial space satellite.

SEA
A flat plain on the Moon, properly called a *mare*.

SOLAR
To do with the Sun.

SOLAR SYSTEM
The family of the Sun, including the planets, moons, asteroids, and comets.

STAR
A huge ball of hot gases that gives off energy as light, heat, and other rays.

SUPERGIANT
The largest kind of star, hundreds of times bigger across than the Sun.

SUPERNOVA
A fantastic explosion that happens when a supergiant star blows itself to bits.

TERRESTRIAL PLANETS
The planets made up mainly of rock: Earth, Mercury, Venus, and Mars.

TRILLION
One million million, or 1,000,000,000,000.

UNIVERSE
All that exists—space and everything in it.

WHITE DWARF
A small, dense star about the same size as Earth.

ZODIAC
An imaginary band in the heavens through which the Sun, Moon, and planets appear to travel. Hence, the constellations of the zodiac.

INDEX

Figures in italics indicate illustrations.

3C-273 quasar 49

A
Achernar 37, 57, 67, 73, 85, 87, 88
Albireo 68
Aldebaran 30, 36, 40, 45, *45*, 72, 82, 87, 89, 90
Aldrin, "Buzz" 54
Algol 91
Alpha Andromedae 79
Alpha Centauri 26, 31, 37, 38, 41, 45, *45*, 51, 53, 57, 89, 90
Alphard 48
Altair 60, 68, 69, 82
Andromeda Galaxy (M31) 76, 78, 80, 81, *81*, 83
Andromeda Nebula 78
Antares 59, 60, 61
Apollo astronauts 17, *17*
April skies 46–49
Aquarius 74, 75
Aquila 50, 60, 68, 69
Ara 62
Arcturus 46, 48, 50, 52, 66, 72
Argo Navis 39, 42, 43, *43*
Aries 84
Armstrong, Neil 17
asteroids 12, *12*
astrology 54–55
astronomy, definition 6, 54
astrophotography 11, *11*
atmosphere 11, 16, 70
atoms 27
August skies 66–71
Auriga 32, *32*
autumnal (fall) equinox 72

B
basalt 17
Beta Centauri 31, 41, 45, *45*, 51, 53, 57, 83, 89
Betelgeuse 31, 33, 36
Big Dipper (Plow) 33, 36, 43, 44, 46, 48, 50, 52, 53, 56, 66, 72, 90
binary stars 32, 68, 88, 91
binoculars 10, *10*
birth chart 55
black holes 27, *53*, 81
blazars 81
Boötes 46, 48, 49, 52, 58
Bopp, Thomas 23

C
cameras 11, *11*
Cancer 36, 38
Canes Venatici 48
Canis Major 32, 33, 37, 38, 86, 88
Canis Minor 33, 38
Canopus 37, 41, 42, 47, 51, 73, 83, 87
Capella 30, 32, *32*, 36, 40, 60, 66, 72, 76, 86
Capricornus 54, 74
Carina (Keel) 39, 42, 43, 45
Cassiopeia 33, 40, 46, 50, 56, 66, 82
Castor 30, 36, 37, 39, *39*, 76
Cat's Eye Nebula *35*
celestial sphere 25, *25*, 28
Centaurus 51, 52–53, 62, *62*
Centaurus A 53, *53*
Cepheids 53, 58
Cepheus 33, 40, 58
Cetus 24, 74, 82, 84
clothing for star-gazing 8, *8*
clusters 65, 90–91
Coal Sack 34
Columba 88
Coma cluster 81
Comet Hale-Bopp 22, *22*, 23
Comet Hyakutake 23
comets 6, 8, 12, 22–23, 70
compass 8
constellations 6, 8, 11, 24, *24*, *25*, 28
Copernicus, Nicolaus 13, *13*
corona *13*
Corona Borealis 58
Corvus 48
craters 16, 71, *71*
Cygnus 24, 44, 50, 57, 60, 68, *69*, 82, 86
Cygnus Rift 66

D
December skies 86–91
Delphinus 68
Deneb (Northern Cross) 44, 46, 50, 56, 57, 60, 68, 76, 82, 86
"dirty snowball" 23
Dorado 74
double stars 32
Draco 33, *35*, 42
Dubhe 44
dust rings 47

E
Eagle Nebula 63
Earth *19*
  color of *20*
  conditions on 20
  orbit round the Sun 12, *12*, 13, 28
eclipses 6, *8*
Equator 25, 40
Eridanus 82, 85, 88
Eta Carinae 42
Europea (a Jupiter moon) 21
evening star 18, *18*
eyepiece 10, *10*, 11

F
falling stars 8, 70
False Cross 31, 42, 43, 45
February skies 36–39
Fomalhaut 57, 67, 73, 75, 76, 87

G
galaxies 6, 10, 49, 80–81
Galileo Galilei 10
Galle, Johann 19
gas giants 21, *21*
Gemini 36, 39, 76, 86
Gemma 58
Giotto probe *23*
globular clusters 91, *91*
gravity 16, 27, 70
Grus 75

H
Hale, Alan 23
Halley, Edmond 22
Halley's Comet 22
helium 27
Hercules 58–59, *91*
Herschel, William 19
Horsehead Nebula 34
Hubble, Edwin 83
Hubble Space Telescope *10*, 11, *27*, 49, *80*
Hyades 82, 87, 89
Hydra 36, 46, 48, 50
hydrogen 27

I
Ida *12*

J
January skies 30–35
Jewel Box star cluster 85
July skies 60–65
June skies 56–59
Jupiter *12*, 18, *18*, 19, *19*, 21, *21*

K
Keck telescopes, Hawaii 11
Kitt Peak National Observatory, Arizona *10*

L
Lagoon nebula 69
Lambda 68
Large Magellanic Cloud (LMC) 31, 74, 75, *75*, 80, 81, *91*
lenses 10
Leo 24, 40, 43, 44, 46, 54, *54*, 70
Leonid shower 70
Lepus 33
Libra 50, 52
light-years 26
Little Dipper 44, 53, 56, 86
Local Group 81
lunatics 15
Lupus 62
Lyra 47, 60, 63

M
M numbers 35
M3 48
M4 59, *59*
M5 63
M6 59
M7 59
M10 63
M11 (Wild Duck) 68
M12 63
M13 59, *91*
M15 79
M16 63
M31 (Andromeda Galaxy) 76, 78, 80, 81, *81*, 83
M33 (Triangulum Galaxy) 81, 85
M42 35
M57 *62*

INDEX

M87 49
M100 *81*
March skies 40–45
mare (plural maria) 15, 16
Mars *12*, 18, *18*, 19, *19*, 20, *20*
May skies 50–55
Megrez 44
Merak 44
Mercury *12*, 18, *19*, 20
Messier, Charles 35
meteorites 16, 17, 70, 71, *71*
meteors 67, 70–71
Milky Way (main reference) 64–65
miniplanets 12, *12*
Mira (Omicron) 84–85
mirrors 11
Mizar 90
Monoceros 38, *38*
Moon
  footsteps on 17
  gravity 16
  light and dark 15
  lunar landscapes 16
  orbit around the Earth 12
  phases of 14–15, *14*
  size of 16
  soil and rocks 17, *17*
moons 12
morning star 18, *18*
Mu (Garnet Star) 58, 88

N
nebulas (gas clouds) 10, 26, *26*, 33, 34–5, 43
Neptune *12*, 19, *19*, 21
night vision 9
North America Nebula 68, 69
North Star *see* Pole Star
Northern Cross *see* Deneb
Northern Hemisphere 24, 25, *25*, 29, *29*
November skies 82–85

O
October skies 76–81
Omega Centauri 51, 53
Omega Nebula 69
Ophiuchus 63
Orion 30, 31, 33, 39, 44–5, 65, 83, 86, 90
Orion Nebula 33, 34, 35, *35*
Orion's Belt 45, *45*

P
Pathfinder *20*
Pavo 59, 75
Pegasus 72, 76, 77, 79, 82
Perseus 24, 65–68, 88, *89*, 91
Phekda 44
Phoenix 59, 85
Pisces 54, 74, 79, 82
Piscis Austrinus 74–75
planetary nebulas *27*, 34, *35*
planets
  bright giants 19
  family of 12
  gas giants 21
  of the night 18
  orbit around the Sun 12, *12*, 13, *13*
  rocky 20, *20*
  sizes 19
  wandering of 6, 8
planisphere 9, *9*
Pleiades 45, *45*, 82, 84, 89, 90–91, *90*
Pluto *12*, 19, *19*, 20, 21
Pole Star (Polaris) 8, 33, 40, 41, 42, 44, 46, 53, 56, 60, 66, 86
Pollux 30, 36, 37, 39, *39*, 76
Praesepe (M44) 38
precession 41
Procyon 30, 36
Proxima Centauri 90
Pup 32
Puppis (Poop) 39

Q
quasars 49, *49*, 81

R
radio galaxies 53, *53*, 81
radio waves 81
red giants 26
red light 9
reflector 11
reflectors *10*
refractors 10
regolith 17
Regulus 44, 46, 86
Rigel 30, 33, 36, 77
Rosette Nebula 38

S
Sagitta 69
Sagittarius 51, 60, 61, 65, *65*, 69, 73, 83
satellites *64*
Saturn *12*, 18, 19, *19*, 21, *21*
Schmitt, Jack *17*
Scorpius 51, 52, 54, 59–62, *59*, 73, 83
sea (flat plain on the Moon) 15, 16
September skies 72–75
Serpens 50, 63
Sirius (Dog Star) 30, 32, 36, 37, 38, 42, 45, 47, 77, 83, 86
Smaller Magellanic Cloud (SMC) 31, 75, 79, 80, 81
SOHO spacecraft 13
Solar System 12, *12*, 13, 21, 22
Southern Cross (Crux) 31, 34, 41, 42, 45, *45*, 47, 53, 61, 65, 77, 85
Southern Hemisphere 25, *25*, 29, *29*
Southern Triangle (Triangulum Australe) 57, 89
Spica 37, 46, 49, 50
spirals 64
spring (vernal) equinox 40
Square of Pegasus 60, 79, *79*, 84, 85
star clusters 10, 43
star maps 9, *9*, 28, 29
star signs 55
star trails 11, *11*
stars
  birth of 26, *26*, *27*, 34
  brightness and color 26
  death of 26–27, 34
  distance from Earth 26
  falling 8, 70
  patterns 24
Sun
  death of 26, *27*
  eclipse of 13
  energy 13
  planetary orbit round 12, *12*, 13, *13*
  size of 13
  temperature 13
superclusters 81
supergiants 27, 33
supernovas 27, 34

T
Tarantula Nebula 74, *75*, 91
Taurus 39, 54, 65, 89
telescopes 10, *10*, 19
  radio 65
tides 16
Tombaugh, Clyde 19
torch *8*
Triangulum constellation 6, *35*, 85
Triangulum galaxy (M33) 81, 85
Trifid Nebula 69
tripod 11, *11*
Triton (a Neptune moon) 21
Tucana 59, 79

U
ultraviolet rays 13
Universe, boundless 25
Uranus *12*, 19, *19*, 21
Ursa Major (Great Bear) 43, 48, 52, 53
Ursa Minor 33, 43, 44, 53, 56

V
variable stars 33, 88
Vega 47, 56, 60, 63, 76, 82
Vela (Sails) 39, 42, 43, 46
Venus *12*, 18, *18*, *19*, 20, *20*
Virgo 46, 49, 50, 81

W
werewolves 15, *15*
Whirlpool Galaxy (M51) 48
white dwarfs 26, 32, *59*
winter solstice 86

X
X-rays 13, 81

Z
zodiac 39, 52, 54–55, 74, 89

CREDITS

Quarto would like to thank:

Telescope House for their generous loan of props.
Allsorts Model Agency for supplying models (Ria Bushell, Alice Piner, Peter Reynolds, Kyri Sawa, C.J.Thomas) used for photography.
Richard Monkhouse and John Cox for producing the starmaps.

The following images are the copyright of the author:
 (b=Bottom, t=Top, c=Center, l=Left, r=Right)
10br, 11tl, 11tr, 12tr, 13lc, 13tr, 14tr, 14c, 15tr, 16tr, 16bl, 17tl, 17br, 18bl, 18tr, 20tr, 20c, 20br, 21t, 21bc, 22bl, 22tr, 23bl, 23tr, 24r, 26bl, 26tr, 27bl, 27r, 32bl, 35l, 35tr, 35br, 38br, 39tr, 42tr, 43tr, 45bc, 49r, 53t, 54bl, 54tr, 59tc, 59tr, 62r, 64r, 65l, 65r, 69r, 70bl, 71lc, 71rc, 71br, 75r, 78br, 79bl, 80bl, 81tl, 81cr, 84tr, 89tl, 90bl, 90tr, 91tl

All other photographs and illustrations are the copyright of Quarto. While every effort has been made to credit contributors, we apologize if there have been any omissions or errors.